Machine Learning with Scala Quick Start Guide

Leverage popular machine learning algorithms and techniques and implement them in Scala

Md. Rezaul Karim

BIRMINGHAM - MUMBAI

Machine Learning with Scala Quick Start Guide

Commissioning Editor: Amey Varangaonkar
Acquisition Editor: Aditi Gour
Content Development Editor: Roshan Kumar
Technical Editor: Nilesh Sawakhande
Copy Editor: Safis Editing
Project Coordinator: Namrata Swetta
Proofreader: Safis Editing
Indexer: Rekha Nair
Graphics: Alishon Mendonsa
Production Coordinator: Shraddha Falebhai

First published: April 2019

Production reference: 1300419

Published by Packt Publishing Ltd.
Livery Place
35 Livery Street
Birmingham
B3 2PB, UK.

ISBN 978-1-78934-507-0

www.packtpub.com

`mapt.io`

Mapt is an online digital library that gives you full access to over 5,000 books and videos, as well as industry leading tools to help you plan your personal development and advance your career. For more information, please visit our website.

Why subscribe?

- Spend less time learning and more time coding with practical eBooks and Videos from over 4,000 industry professionals

- Improve your learning with Skill Plans built especially for you

- Get a free eBook or video every month

- Mapt is fully searchable

- Copy and paste, print, and bookmark content

Packt.com

Did you know that Packt offers eBook versions of every book published, with PDF and ePub files available? You can upgrade to the eBook version at `www.packt.com` and as a print book customer, you are entitled to a discount on the eBook copy. Get in touch with us at `customercare@packtpub.com` for more details.

At `www.packt.com`, you can also read a collection of free technical articles, sign up for a range of free newsletters, and receive exclusive discounts and offers on Packt books and eBooks.

Contributors

About the author

Md. Rezaul Karim is a researcher, author, and data science enthusiast with a strong computer science background, plus 10 years of R&D experience in machine learning, deep learning, and data mining algorithms to solve emerging bioinformatics research problems by making them explainable. He is passionate about applied machine learning, knowledge graphs, and explainable artificial intelligence (XAI).

Currently, he is working as a research scientist at Fraunhofer FIT, Germany. He is also a Ph.D. candidate at RWTH Aachen University, Germany. Before joining FIT, he worked as a researcher at the Insight Centre for Data Analytics, Ireland. Previously, he worked as a lead software engineer at Samsung Electronics, Korea.

About the reviewers

Ajay Kumar N has experience in big data, and specializes in cloud computing and various big data frameworks, including Apache Spark and Apache Hadoop. His primary language of choice is Python, but he also has a special interest in functional programming languages such as Scala. He has worked extensively with NumPy, pandas, and scikit-learn, and often contributes to open source projects related to data science and machine learning.

Sarbashree Ray has over 5 years' experience in big data analytics, currently at Reliance Jio as a deputy manager. Sarbashree is an engineering professional with experience of designing and executing solutions for complex business problems involving large-scale big data and machine learning technologies, real-time analytics, and reporting solutions. He is also known for using the right tools when and where they make sense, and creating intuitive architectures that help organizations effectively analyze and process terabytes of structured and unstructured data. He is also able to integrate state-of-the-art big data technologies into overall architectures and lead a team of developers through the construction, testing, and implementation phases.

Packt is searching for authors like you

If you're interested in becoming an author for Packt, please visit `authors.packtpub.com` and apply today. We have worked with thousands of developers and tech professionals, just like you, to help them share their insight with the global tech community. You can make a general application, apply for a specific hot topic that we are recruiting an author for, or submit your own idea.

Table of Contents

Preface

Machine learning has made a huge impact not only in academia, but also in industry, by turning data into actionable intelligence. Scala is not only an object-oriented and functional programming language, but can also leverage the advantages of **Java Virtual Machine (JVM)**. Scala provides code complexity optimization and offers concise notation, which is probably the reason it has seen a steady rise in adoption over the last few years, especially in data science and analytics.

This book is aimed at aspiring data scientists, data engineers, and deep learning enthusiasts who are newbies and want to have a great head start at machine learning best practices. Even if you're not well versed in machine learning concepts, but still want to expand your knowledge by delving into practical implementations of supervised learning, unsupervised learning, and recommender systems with Scala, you will be able to grasp the content easily!

Throughout the chapters, you'll become acquainted with popular machine learning libraries in Scala, learning how to carry out regression and classification analysis using both linear methods and tree-based ensemble techniques, as well as looking at clustering analysis, dimensionality reduction, and recommender systems, before delving into deep learning at the end.

After reading this book, you will have a good head start in solving more complex machine learning tasks. This book isn't meant to be read cover to cover. You can turn the pages to a chapter that looks like something you're trying to accomplish or that ignites your interest.

Suggestions for improvement are always welcome. Happy reading!

Who this book is for

Machine learning developers looking to learn how to train machine learning models in Scala, without spending too much time and effort, will find this book to be very useful. Some fundamental knowledge of Scala programming and some basics of statistics and linear algebra is all you need to get started with this book.

What this book covers

Chapter 1, *Introduction to Machine Learning with Scala*, first explains some basic concepts of machine learning and different learning tasks. It then discusses Scala-based machine learning libraries, which is followed by configuring your programming environment.

Finally, it covers Apache Spark briefly, before demonstrating a step-by-step example.

Chapter 2, *Scala for Regression Analysis*, covers a supervised learning task called regression analysis with examples, followed by regression metrics. It then explains some regression analysis algorithms, including linear regression and generalized linear regression. Finally, it demonstrates a step-by-step solution to a regression analysis task using Spark ML in Scala.

Chapter 3, *Scala for Learning Classification*, briefly explains another supervised learning task called classification with examples, followed by explaining how to interpret performance evaluation metrics. It then covers widely used classification algorithms such as logistic regression, Naïve Bayes, and **support vector machines (SVMs)**. Finally, it demonstrates a step-by-step solution to a classification problem using Spark ML in Scala.

Chapter 4, *Scala for Tree-Based Ensemble Techniques*, covers very powerful and widely used tree-based approaches, including decision trees, gradient-boosted trees, and random forest algorithms, for both classification and regression analysis. It then revisits the examples of Chapter 2, *Scala for Regression Analysis*, and Chapter 3, *Scala for Learning Classification*, before solving them using these tree-based algorithms.

Chapter 5, *Scala for Dimensionality Reduction and Clustering*, briefly discusses different clustering analysis algorithms, followed by a step-by-step example of solving a clustering problem. Finally, it discusses the curse of dimensionality in high-dimensional data, before showing an example of solving it using **principal component analysis (PCA)**.

Chapter 6, *Scala for Recommender System*, briefly covers similarity-based, content-based, and collaborative filtering approaches for developing recommendation systems. Finally, it demonstrates an example of a book recommender system with Spark ML in Scala.

Chapter 7, *Introduction to Deep Learning with Scala*, briefly covers deep learning, artificial neural networks, and neural network architectures. It then discusses some available deep learning frameworks. Finally, it demonstrates a step-by-step example of solving a cancer type prediction problem using a **long short-term memory (LSTM)** network.

To get the most out of this book

All the examples have been implemented in Scala with some open source libraries, including Apahe Spark MLlib/ML and Deeplearning4j. However, to get the best out of this, you should have a powerful computer and software stack.

A Linux distribution is preferable (for example, Debian, Ubuntu, or CentOS). For example, for Ubuntu, it is recommended to have at least a 14.04 (LTS) 64-bit complete installation on VMware Workstation Player 12 or VirtualBox. You can run Spark jobs on Windows (7/8/10) or macOS X (10.4.7+) as well.

A computer with a Core i5 processor, enough storage (for example, for running Spark jobs, you'll need at least 50 GB of free disk storage for standalone cluster and for the SQL warehouse), and at least 16 GB RAM are recommended. And optionally, if you want to perform the neural network training on the GPU (for the last chapter only), the NVIDIA GPU driver has to be installed with CUDA and CuDNN configured.

The following APIs and tools are required in order to execute the source code in this book:

- Java/JDK, version 1.8
- Scala, version 2.11.8
- Spark, version 2.2.0 or higher
- Spark csv_2.11, version 1.3.0
- ND4j backend version nd4j-cuda-9.0-platform for GPU; otherwise, nd4j-native
- ND4j, version 1.0.0-alpha
- DL4j, version 1.0.0-alpha
- Datavec, version 1.0.0-alpha
- Arbiter, version 1.0.0-alpha
- Eclipse Mars or Luna (latest version) or IntelliJ IDEA
- Maven Eclipse plugin (2.9 or higher)
- Maven compiler plugin for Eclipse (2.3.2 or higher)
- Maven assembly plugin for Eclipse (2.4.1 or higher)

Download the example code files

You can download the example code files for this book from your account at www.packt.com. If you purchased this book elsewhere, you can visit www.packt.com/support and register to have the files emailed directly to you.

You can download the code files by following these steps:

1. Log in or register at www.packt.com.
2. Select the **SUPPORT** tab.
3. Click on **Code Downloads & Errata**.
4. Enter the name of the book in the **Search** box and follow the onscreen instructions.

Once the file is downloaded, please make sure that you unzip or extract the folder using the latest version of:

- WinRAR/7-Zip for Windows
- Zipeg/iZip/UnRarX for Mac
- 7-Zip/PeaZip for Linux

The code bundle for the book is also hosted on GitHub at `https://github.com/PacktPublishing/Machine-Learning-with-Scala-Quick-Start-G uide`. In case there's an update to the code, it will be updated on the existing GitHub repository.

We also have other code bundles from our rich catalog of books and videos available at `https://github.com/PacktPublishing/`. Check them out!

Code in Action

Visit the following link to check out videos of the code being run:
`http://bit.ly/2WhQf2i`

Conventions used

There are a number of text conventions used throughout this book.

`CodeInText`: Indicates code words in text, database table names, folder names, filenames, file extensions, pathnames, dummy URLs, user input, and Twitter handles. Here is an example: "It gave me a Matthews correlation coefficient of `0.3888239300421191`."

A block of code is set as follows:

```
rawTrafficDF.select("Hour (Coded)", "Immobilized bus", "Broken Truck",
                    "Vehicle excess", "Fire", "Slowness in traffic
(%)").show(5)
```

When we wish to draw your attention to a particular part of a code block, the relevant lines or items are set in bold:

```
// Create a decision tree estimator
val dt = new DecisionTreeClassifier()
    .setImpurity("gini")
    .setMaxBins(10)
    .setMaxDepth(30)
    .setLabelCol("label")
    .setFeaturesCol("features")
```

Any command-line input or output is written as follows:

```
+-----+-----+
|churn|count|
+-----+-----+
|False| 2278|
| True| 388 |
+-----+-----+
```

Bold: Indicates a new term, an important word, or words that you see onscreen. For example, words in menus or dialog boxes appear in the text like this. Here is an example: "Clicking the **Next** button moves you to the next screen."

 Warnings or important notes appear like this.

 Tips and tricks appear like this.

Get in touch

Feedback from our readers is always welcome.

General feedback: If you have questions about any aspect of this book, mention the book title in the subject of your message and email us at customercare@packtpub.com.

Errata: Although we have taken every care to ensure the accuracy of our content, mistakes do happen. If you have found a mistake in this book, we would be grateful if you would report this to us. Please visit www.packt.com/submit-errata, selecting your book, clicking on the Errata Submission Form link, and entering the details.

Piracy: If you come across any illegal copies of our works in any form on the Internet, we would be grateful if you would provide us with the location address or website name. Please contact us at copyright@packt.com with a link to the material.

If you are interested in becoming an author: If there is a topic that you have expertise in and you are interested in either writing or contributing to a book, please visit authors.packtpub.com.

Reviews

Please leave a review. Once you have read and used this book, why not leave a review on the site that you purchased it from? Potential readers can then see and use your unbiased opinion to make purchase decisions, we at Packt can understand what you think about our products, and our authors can see your feedback on their book. Thank you!

For more information about Packt, please visit `packt.com`.

1
Introduction to Machine Learning with Scala

In this chapter, we will explain some basic concepts of **machine learning** (**ML**) that will be used in all subsequent chapters. We will start with a brief introduction to ML including basic learning workflow, ML rule of thumb, and different learning tasks. Then we will gradually cover most important ML tasks.

Also, we will discuss getting started with Scala and Scala-based ML libraries for getting a quick start for the next chapter. Finally, we get started with ML with Scala and Spark ML by solving a real-life problem. The chapter will briefly cover the following topics:

- Overview of ML
- ML tasks
- Introduction to Scala
- Scala ML libraries
- Getting started with ML with Spark ML

Technical requirements

You'll be required to have basic knowledge of Scala and Java. Since Scala is also a JVM-based language, make sure both Java JRE and JDK are installed and configured on your machine. To be more specific, you'll need Scala 2.11.x and Java 1.8.x version installed. Also, you need an IDE, such as Eclipse, IntelliJ IDEA, or Scala IDE, with the necessary plugins. However, if you're using IntelliJ IDEA, Scala will already be integrated.

The code files of this chapter can be found on GitHub:

```
https://github.com/PacktPublishing/Machine-Learning-with-Scala-Quick-Start-
Guide/tree/master/Chapter01
```

Check out the following video to see the Code in Action:
```
http://bit.ly/2V3Id08
```

Overview of ML

ML approaches are based on a set of statistical and mathematical algorithms in order to carry out tasks such as classification, regression analysis, concept learning, predictive modeling, clustering, and mining of useful patterns. Using ML, we aim to improve the whole learning process automatically such that we may not need complete human interactions, or we can at least reduce the level of such interactions as much as possible.

Working principles of a learning algorithm

Tom M. Mitchell explained what learning really means from a computer science perspective:

> *"A computer program is said to learn from experience E with respect to some class of tasks T and performance measure P, if its performance at tasks in T, as measured by P, improves with experience E."*

Based on this definition, we can conclude that a computer program or machine can do the following:

- Learn from data and histories
- Improve with experience
- Iteratively enhance a model that can be used to predict outcomes of questions

Since the preceding points are at the core of predictive analytics, almost every ML algorithm we use can be treated as an optimization problem. This is about finding parameters that minimize an objective function, for example, a weighted sum of two terms such as a cost function and regularization. Typically, an objective function has two components:

- A regularizer, which controls the complexity of the model
- The loss, which measures the error of the model on the training data

On the other hand, the regularization parameter defines the trade-off between minimizing the training error and the model's complexity, in an effort to avoid overfitting problems. Now, if both of these components are convex, then their sum is also convex. So, when using an ML algorithm, the goal is to obtain the best hyperparameters of a function that return the minimum error when making predictions. Therefore, by using a convex optimization technique, we can minimize the function until it converges toward the minimum error.

Given that a problem is convex, it is usually easier to analyze the asymptotic behavior of the algorithm, which shows how fast it converges as the model observes more and more training data. The task of ML is to train a model so that it can recognize complex patterns from the given input data and can make decisions in an automated way.

Thus, inferencing is all about testing the model against new (that is, unobserved) data and evaluating the performance of the model itself. However, in the whole process and for making the predictive model a successful one, data acts as the first-class citizen in all ML tasks. In reality, the data that we feed to our machine learning systems must be made up of mathematical objects, such as vectors, so that they can consume such data. For example, in the following diagram, raw images are embedded into numeric values called feature vectors before feeding in to the learning algorithm:

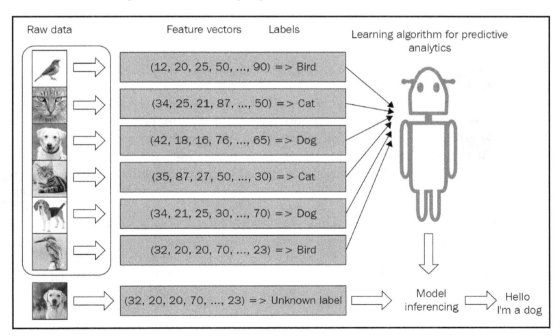

Depending on the available data and feature types, the performance of your predictive model can vacillate dramatically. Therefore, selecting the right features is one of the most important steps before the inferencing takes place. This is called feature engineering, where the domain knowledge about the data is used to create only selective or useful features that help prepare the feature vectors to be used so that a machine learning algorithm works.

For example, comparing hotels is quite difficult unless we already have a personal experience of staying in multiple hotels. However, with the help of an ML model, which is already trained with quality features out of thousands of reviews and features (for example, how many stars does a hotel have, size of the room, location, room service, and so on), it is pretty feasible now. We'll see several examples throughout the chapters. However, before developing such an ML model, knowing some ML concepts is also important.

General machine learning rule of thumb

The general machine learning rule of thumb is that the more data there is, the better the predictive model. However, having more features often creates a mess, to the extent that the performance degrades drastically, especially if the dataset is high-dimensional. The entire learning process requires input datasets that can be split into three types (or are already provided as such):

- A **training set** is the knowledge base coming from historical or live data that is used to fit the parameters of the ML algorithm. During the training phase, the ML model utilizes the training set to find optimal weights of the network and reach the objective function by minimizing the training error. Here, the back-prop rule or an optimization algorithm is used to train the model, but all the hyperparameters are needed to be set before the learning process starts.
- A **validation set** is a set of examples used to tune the parameters of an ML model. It ensures that the model is trained well and generalizes toward avoiding overfitting. Some ML practitioners refer to it as a development set or dev set as well.
- A **test set** is used for evaluating the performance of the trained model on unseen data. This step is also referred to as model inferencing. After assessing the final model on the test set (that is, when we're fully satisfied with the model's performance), we do not have to tune the model any further, but the trained model can be deployed in a production-ready environment.

A common practice is splitting the input data (after necessary pre-processing and feature engineering) into 60% for training, 10% for validation, and 20% for testing, but it really depends on use cases. Sometimes, we also need to perform up-sampling or down-sampling on the data based on the availability and quality of the datasets.

This rule of thumb of learning on different types of training sets can differ across machine learning tasks, as we will cover in the next section. However, before that, let's take a quick look at a few common phenomena in machine learning.

General issues in machine learning models

When we use this input data for the training, validation, and testing, usually the learning algorithms cannot learn 100% accurately, which involves training, validation, and test error (or loss). There are two types of error that one can encounter in a machine learning model:

- Irreducible error
- Reducible error

The irreducible error cannot be reduced even with the most robust and sophisticated model. However, the reducible error, which has two components, called bias and variance, can be reduced. Therefore, to understand the model (that is, prediction errors), we need to focus on bias and variance only:

- Bias means how far the predicted value are from the actual values. Usually, if the average predicted values are very different from the actual values (labels), then the bias is higher.
- An ML model will have a high bias because it can't model the relationship between input and output variables (can't capture the complexity of data well) and becomes very simple. Thus, a too-simple model with high variance causes underfitting of the data.

The following diagram gives some high-level insights and also shows what a just-right fit model should look like:

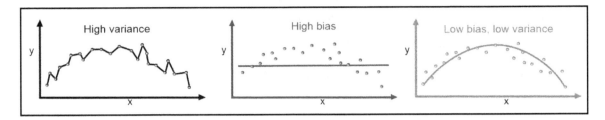

Variance signifies the variability between the predicted values and the actual values (how scattered they are).

Identifying high bias and high variance: If the model has a high training error as well as the validation error or test error is the same as the training error, the model has high bias. On the other hand, if the model has low training error but has high validation or high test error, the model has a high variance.

An ML model usually performs very well on the training set but doesn't work well on the test set (because of high error rates). Ultimately, it results in an underfit model. We can recap the overfitting and underfitting once more:

- **Underfitting**: If your training and validation error are both relatively equal and very high, then your model is most likely underfitting your training data.
- **Overfitting**: If your training error is low and your validation error is high, then your model is most likely overfitting your training data. The just-rightfit model learns very well and performs better on unseen data too.

Bias-variance trade-off: The high bias and high variance issue is often called bias-variance trade-off, because a model cannot be too complex or too simple at the same time. Ideally, we would strive for the best model that has both low bias and low variance.

Now we know the basic working principle of an ML algorithm. However, based on problem type and the method used to solve a problem, ML tasks can be different, for example, supervised learning, unsupervised learning, and reinforcement learning. We'll discuss these learning tasks in more detail in the next section.

ML tasks

Although every ML problem is more or less an optimization problem, the way they are solved can vary. In fact, learning tasks can be categorized into three types: supervised learning, unsupervised learning, and reinforcement learning.

Supervised learning

Supervised learning is the simplest and most well-known automatic learning task. It is based on a number of predefined examples, in which the category to which each of the inputs should belong is already known, as shown in the following diagram:

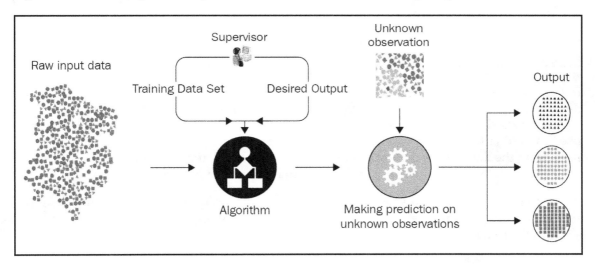

The preceding diagram shows a typical workflow of supervised learning. An actor (for example, a data scientist or data engineer) performs **Extraction Transformation Load (ETL)** and the necessary feature engineering (including feature extraction, selection, and so on) to get the appropriate data with features and labels so that they can be fed in to the model. Then he would split the data into training, development, and test sets. The training set is used to train an ML model, the validation set is used to validate the training against the overfitting problem and regularization, and then the actor would evaluate the model's performance on the test set (that is, unseen data).

However, if the performance is not satisfactory, he can perform additional tuning to get the best model based on hyperparameter optimization. Finally, he would deploy the best model in a production-ready environment. The following diagram summarizes these steps in a nutshell:

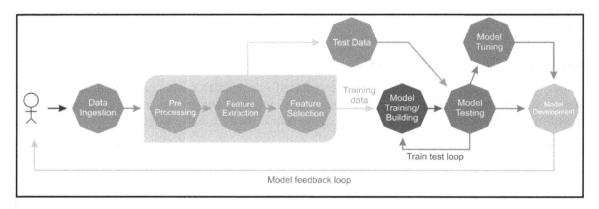

In the overall life cycle, there might be many actors involved (for example, a data engineer, data scientist, or an ML engineer) to perform each step independently or collaboratively. The supervised learning context includes classification and regression tasks; classification is used to predict which class a data point is a part of (discrete value). It is also used for predicting the label of the class attribute. On the other hand, regression is used for predicting continuous values and making a numeric prediction of the class attribute.

In the context of supervised learning, the learning process required for the input dataset is split randomly into three sets, for example, 60% for the training set, 10% for the validation set, and the remaining 30% for the testing set.

Unsupervised learning

How would you summarize and group a dataset if the labels were not given? Probably, you'll try to answer this question by finding the underlying structure of a dataset and measuring the statistical properties such as frequency distribution, mean, standard deviation, and so on. If the question is *how would you effectively represent data in a compressed format?* You'll probably reply saying that you'll use some software for doing the compression, although you might have no idea how that software would do it. The following diagram shows the typical workflow of an unsupervised learning task:

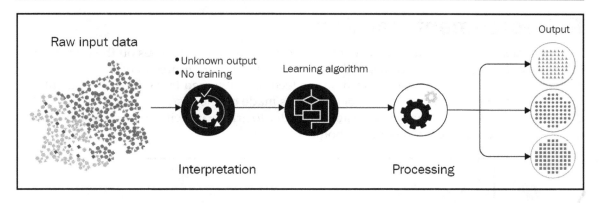

These are exactly two of the main goals of unsupervised learning, which is largely a data-driven process. We call this type of learning *unsupervised* because you will have to deal with unlabeled data. The following quote comes from Yann LeCun, director of AI research (source: Predictive Learning, NIPS 2016, Yann LeCun, Facebook Research):

> *"Most of human and animal learning is unsupervised learning. If intelligence was a cake, unsupervised learning would be the cake, supervised learning would be the icing on the cake, and reinforcement learning would be the cherry on the cake. We know how to make the icing and the cherry, but we don't know how to make the cake. We need to solve the unsupervised learning problem before we can even think of getting to true AI".*

The two most widely used unsupervised learning tasks include the following:

- **Clustering**: Grouping data points based on similarity (or statistical properties). For example, a company such as Airbnb often groups its apartments and houses into neighborhoods so that customers can navigate the listed ones more easily.
- **Dimensionality reduction**: Compressing the data with the structure and statistical properties preserved as much as possible. For example, often the number of dimensions of the dataset needs to be reduced for the modeling and visualization.
- **Anomaly detection**: Useful in several applications such as identification of credit card fraud detection, identifying faulty pieces of hardware in an industrial engineering process, and identifying outliers in large-scale datasets.
- **Association rule mining**: Often used in market basket analysis, for example, asking which items are brought together and frequently.

Reinforcement learning

Reinforcement learning is an artificial intelligence approach that focuses on the learning of the system through its interactions with the environment. In reinforcement learning, the system's parameters are adapted based on the feedback obtained from the environment, which in turn provides feedback on the decisions made by the system. The following diagram shows a person making decisions in order to arrive at their destination. Let's take an example of the route you take from home to work:

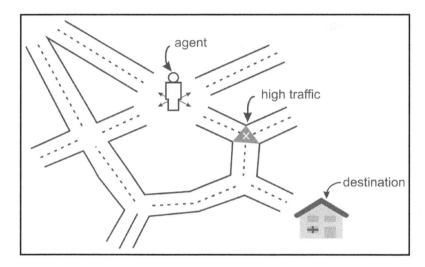

In this case, you take the same route to work every day. However, out of the blue, one day you get curious and decide to try a different route with a view to finding the shortest path. Similarly, based on your experience and the time taken with the different route, you'd decide whether you should take a specific route more often. We can take a look at one more example in terms of a system modeling a chess player. In order to improve its performance, the system utilizes the result of its previous moves; such a system is said to be a system learning with reinforcement.

So far, we have learned the basic working principles of ML and different learning tasks. However, a summarized view of each learning task with some example use cases is a mandate, which we will see in the next subsection.

Summarizing learning types with applications

We have seen the basic working principles of ML algorithms. Then we have seen what the basic ML tasks are and how they formulate domain-specific problems. However, each of these learning tasks can be solved using different algorithms. The following diagram provides a glimpse into this:

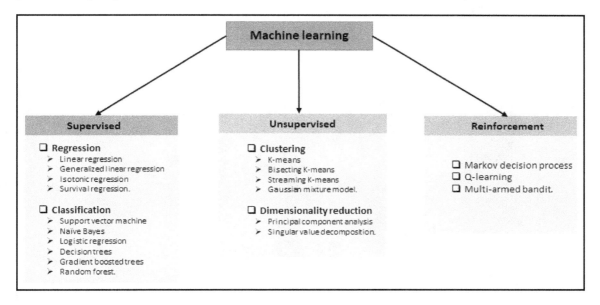

Types of learning and related problems

The following diagram summarizes the previously mentioned ML tasks and some applications:

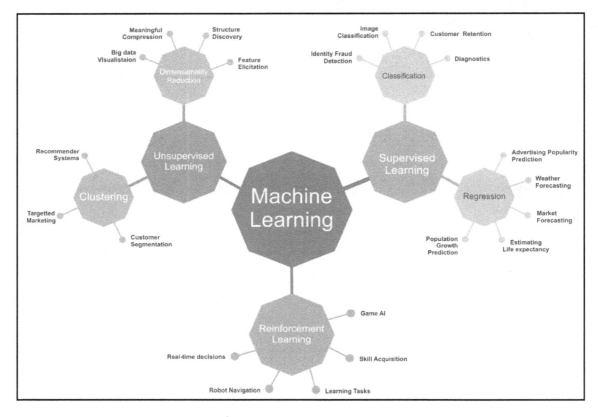

ML tasks and some use cases from different application domains

However, the preceding diagram lists only a few use cases and applications using different ML tasks. In practice, ML is used in numerous use cases and applications. We will try to cover a few of those throughout this book.

Overview of Scala

Scala is a scalable, functional, and object-oriented programming language that is most closely related to Java. However, Scala is designed to be more concise and have features of functional programming languages. For example, Apache Spark, which is written in Scala, is a fast and general engine for large-scale data processing.

Scala's success is due to many factors: it has many tools that enable succinct expression, it is very concise because you need less typing, and it therefore requires less reading, and it offers very good performance as well. This is why Spark has more support for Scala in the sense that more APIs are available that are written in Scala compared to R, Python, and Java. Scala's symbolic operators are easy to read and, compared to Java, most of the Scala codes are comparatively concise and easy to read; Java is too verbose. Functional programming concepts such as pattern matching and higher-order functions are also available in Scala.

The best way to get started with Scala is either using Scala through the **Scala build tool (SBT)** or to use Scala through an **integrated development environment (IDE)**. Either way, the first important step is downloading, installing, and configuring Scala. However, since Scala runs on **Java Virtual Machine (JVM)**, having Java installed and configured on your machine is a prerequisite. Therefore, I'm not going to cover how to do that. Instead, I will provide some useful links (`https://en.wikipedia.org/wiki/Integrated_development_environment`).

Just follow the instructions on how to set up both Java and an IDE (for example, IntelliJ IDEA) or build tool (for example, SBT) at `https://www.scala-lang.org/download/`. If you're using Windows (for example, Windows 10) or Linux (for example, Ubuntu), visit `https://www.journaldev.com/7456/download-install-scala-linux-unix-windows`. Finally, here are some macOS instructions: `http://sourabhbajaj.com/mac-setup/Scala/README.html`.

Java programmers normally prefer Scala when they need to add some functional programming flavor to their codes as Scala runs on JVM. There are various other options when it comes to editors. The following are some options to choose from:

- Scala IDE
- Scala plugin for Eclipse
- IntelliJ IDEA
- Emacs
- Vim

Eclipse has several advantages using numerous beta plugins and local, remote, and high-level debugging facilities with semantic highlighting and code completion for Scala.

ML libraries in Scala

Although Scala is a relatively new programming language compared to Java and Python, the question will arise as to why we need to consider learning it while we have Python and R. Well, Python and R are two leading programming languages for rapid prototyping and data analytics including building, exploring, and manipulating powerful models.

But Scala is becoming the key language too in the development of functional products, which are well suited for big data analytics. Big data applications often require stability, flexibility, high speed, scalability, and concurrency. All of these requirements can be fulfilled with Scala because Scala is not only a general-purpose language but also a powerful choice for data science (for example, Spark MLlib/ML). I've been using Scala for the last couple of years and I found that more and more Scala ML libraries are in development. Up next, we will discuss available and widely used Scala libraries that can be used for developing ML applications.

Interested readers can take a quick look at this, which lists the 15 most popular Scala libraries for ML and data science:
`https://www.datasciencecentral.com/profiles/blogs/top-15-scala-libraries-for-data-science-in-2018-1`

Spark MLlib and ML

MLlib is a library that provides user-friendly ML algorithms that are implemented using Scala. The same API is then exposed to provide support for other languages such as Java, Python, and R. Spark MLlib provides support for local vectors and matrix data types stored on a single machine, as well as distributed matrices backed by one or multiple **resilient distributed datasets** (**RDDs**).

RDD is the primary data abstraction of Apache Spark, often called Spark Core, that represents an immutable, partitioned collection of elements that can be operated on in parallel. The resiliency makes RDD fault-tolerant (based on RDD lineage graph). RDD can help in distributed computing even when data is stored on multiple nodes in a Spark cluster. Also, RDD can be converted into a dataset as a collection of partitioned data with primitive values such as tuples or other objects.

Spark ML is a new set of ML APIs that allows users to quickly assemble and configure practical machine learning pipelines on top of datasets, which makes it easier to combine multiple algorithms into a single pipeline. For example, an ML algorithm (called estimator) and a set of transformers (for example, a `StringIndexer`, a `StandardScalar`, and a `VectorAssembler`) can be chained together to perform the ML task as stages without needing to run them sequentially.

 Interested readers can take a look at the Spark MLlib and ML guide at `https://spark.apache.org/docs/latest/ml-guide.html`.

At this point, I have to inform you of something very useful. Since we will be using Spark MLlib and ML APIs in upcoming chapters too. Therefore, it would be worth fixing some issues in advance. If you're a Windows user, then let me tell you about a very weird issue that you will experience while working with Spark. The thing is that Spark works on Windows, macOS, and Linux. While using Eclipse or IntelliJ IDEA to develop your Spark applications on Windows, you might face an I/O exception error and, consequently, your application might not compile successfully or may be interrupted.

Spark needs a runtime environment for Hadoop on Windows too. Unfortunately, the binary distribution of Spark (v2.4.0, for example) does not contain Windows-native components such as `winutils.exe` or `hadoop.dll`. However, these are required (not optional) to run Hadoop on Windows if you cannot ensure the runtime environment, an I/O exception saying the following will appear:

```
03/02/2019 11:11:10 ERROR util.Shell: Failed to locate the winutils binary
in the hadoop binary path
  java.io.IOException: Could not locate executable null\bin\winutils.exe in
the Hadoop binaries.
```

There are two ways to tackle this issue on Windows and from IDEs such as Eclipse and IntelliJ IDEA:

1. Download `winutls.exe` from `https://github.com/steveloughran/ winutils/ tree/ master/hadoop-2. 7. 1/bin/`.
2. Download and copy it inside the `bin` folder in the Spark distribution—for example, `spark-2.2.0-bin-hadoop2.7/bin/`.
3. Select **Project** | **Run Configurations...** | **Environment** | **New** | and create a variable named `HADOOP_HOME`, then put the path in the **Value** field. Here is an example: `c:/spark-2.2.0-bin-hadoop2.7/bin/` | **OK** | **Apply** | **Run**.

ScalNet and DynaML

ScalNet is a wrapper around Deeplearning4J intended to emulate a Keras-like API for developing deep learning applications. If you're already familiar with neural network architectures and are coming from a JVM background, it would be worth exploring the Scala-based ScalNet library:

- GitHub (`https://github.com/deeplear.../deeplearning4j/.../master/scalnet`)
- Example (`https://github.com/.../sc.../org/deeplearning4j/scalnet/examples`)

DynaML is a Scala and JVM ML toolbox for research, education, and industry. This library provides an interactive, end-to-end, and enterprise-friendly way of developing ML applications. If you're interested, see more at `https://transcendent-ai-labs.github.io/DynaML/`.

ScalaNLP, Vegas, and Breeze

Breeze is one of the primary scientific computing libraries for Scala, which provides a fast and efficient way of data manipulation operations such as matrix and vector operations for creating, transposing, filling with numbers, conducting element-wise operations, and calculating determinants.

Breeze enables basic operations based on the `netlib-java` library, which enables extremely fast algebraic computations. In addition, Breeze provides a way to perform signal-processing operations, necessary for working with digital signals.

The following are the GitHub links:

- Breeze (`https://github.com/scalanlp/breeze/`)
- Breeze examples (`https://github.com/scalanlp/breeze-examples`)
- Breeze quickstart (`https://github.com/scalanlp/breeze/wiki/Quickstart`)

On the other hand, ScalaNLP is a suite of scientific computing, ML, and natural language processing, which also acts as an umbrella project for several libraries, including Breeze and Epic. Vegas is another Scala library for data visualization, which allows plotting specifications such as filtering, transformations, and aggregations. Vegas is more functional than the other numerical processing library, Breeze.

For more information and examples of using Vegas and Breeze, refer to GitHub:

- Vegas (`https://github.com/vegas-viz/Vegas`)
- Breeze (`https://github.com/scalanlp/breeze`)

Whereas the visualization library of Breeze is backed by Breeze and JFreeChart, Vegas can be considered a missing Matplotlib for Scala and Spark, because it provides several options for rendering plots through and within interactive notebook environments, such as Jupyter and Zeppelin.

Refer to Zeppelin notebook solutions of each chapter in the GitHub repository of this book.

Getting started learning

In this section, we'll see a real-life example of a classification problem. The idea is to develop a classifier that, given the values for sex, age, time, number of warts, type, and area, will predict whether a patient has to go through the cryotherapy.

Description of the dataset

We will use a recently added cryotherapy dataset from the UCI machine learning repository. The dataset can be downloaded from `http://archive.ics.uci.edu/ml/datasets/Cryotherapy+Dataset+#`.

This dataset contains information about wart treatment results of 90 patients using cryotherapy. In case you don't know, a wart is a kind of skin problem caused by infection with a type of human papillomavirus. Warts are typically small, rough, and hard growths that are similar in color to the rest of the skin.

There are two available treatments for this problem:

- **Salicylic acid**: A type of gel containing salicylic acid used in medicated band-aids.
- **Cryotherapy**: A freezing liquid (usually nitrogen) is sprayed onto the wart. It will destroy the cells in the affected area. After the cryotherapy, usually, a blister develops, which eventually turns into a scab and falls off after a week or so.

There are 90 samples or instances that were either recommended to go through cryotherapy or be discharged without cryotherapy. There are seven attributes in the dataset:

- `sex`: Patient gender, characterized by 1 (male) or 0 (female).
- `age`: Patient age.
- `Time`: Observation and treatment time in hours.
- `Number_of_Warts`: Number of warts.
- `Type`: Types of warts.
- `Area`: The amount of affected area.
- `Result_of_Treatment`: The recommended result of the treatment, characterized by either 1 (yes) or 0 (no). It is also the target column.

As you can understand, it is a classification problem because we will have to predict discrete labels. More specifically, it is a binary classification problem. Since this is a small dataset with only six features, we can start with a very basic classification algorithm called logistic regression, where the logistic function is applied to the regression to get the probabilities of it belonging in either class. We will learn more details about logistic regression and other classification algorithms in `Chapter 3`, *Scala for Learning Classification*. For this, we use the Spark ML-based implementation of logistic regression in Scala.

Configuring the programming environment

I am assuming that Java is already installed on your machine and `JAVA_HOME` is set too. Also, I'm assuming that your IDE has the Maven plugin installed. If so, then just create a Maven project and add the project properties as follows:

```
<properties>
    <project.build.sourceEncoding>UTF-8</project.build.sourceEncoding>
    <java.version>1.8</java.version>
    <jdk.version>1.8</jdk.version>
    <spark.version>2.3.0</spark.version>
</properties>
```

In the preceding `properties` tag, I specified the Spark version (that is, `2.3.0`), but you can adjust it. Then add the following dependencies in the `pom.xml` file:

```xml
<dependencies>
    <dependency>
        <groupId>org.apache.spark</groupId>
        <artifactId>spark-core_2.11</artifactId>
        <version>${spark.version}</version>
    </dependency>
    <dependency>
        <groupId>org.apache.spark</groupId>
        <artifactId>spark-sql_2.11</artifactId>
        <version>${spark.version}</version>
    </dependency>
    <dependency>
        <groupId>org.apache.spark</groupId>
        <artifactId>spark-mllib_2.11</artifactId>
        <version>${spark.version}</version>
    </dependency>
    <dependency>
        <groupId>org.apache.spark</groupId>
        <artifactId>spark-graphx_2.11</artifactId>
        <version>${spark.version}</version>
    </dependency>
    <dependency>
        <groupId>org.apache.spark</groupId>
        <artifactId>spark-yarn_2.11</artifactId>
        <version>${spark.version}</version>
    </dependency>
    <dependency>
        <groupId>org.apache.spark</groupId>
        <artifactId>spark-network-shuffle_2.11</artifactId>
        <version>${spark.version}</version>
    </dependency>
    <dependency>
        <groupId>org.apache.spark</groupId>
        <artifactId>spark-streaming-flume_2.11</artifactId>
        <version>${spark.version}</version>
    </dependency>
    <dependency>
        <groupId>com.databricks</groupId>
        <artifactId>spark-csv_2.11</artifactId>
        <version>1.3.0</version>
    </dependency>
</dependencies>
```

Then, if everything goes smoothly, all the JAR files will be downloaded in the project home as Maven dependencies. Alright! Then we can start writing the code.

Getting started with Apache Spark

Since you're here to learn how to solve a real-life problem in Scala, exploring available Scala libraries would be worthwhile. Unfortunately, we don't have many options except for the Spark MLlib and ML, which can be used for the regression analysis very easily and comfortably. Importantly, it has every regression analysis algorithm implemented as high-level interfaces. I assume that Scala, Java, and your favorite IDE such as Eclipse or IntelliJ IDEA are already configured on your machine. We will introduce some concepts of Spark without providing much detail, but we will continue learning in upcoming chapters too.

First, I'll introduce SparkSession, which is a unified entry point of a Spark application introduced from Spark 2.0. Technically, SparkSession is the gateway to interact with some of Spark's functionality with a few constructs such as SparkContext, HiveContext, and SQLContext, which are all encapsulated in a SparkSession. Previously, you have seen how to create such a session, probably without knowing it. Well, a SparkSession can be created as a builder pattern as follows:

```
import org.apache.spark.sql.SparkSession
val spark = SparkSession
        .builder // the builder itself
        .master("local[4]") // number of cores (i.e. 4, use * for all cores)
        .config("spark.sql.warehouse.dir", "/temp") // Spark SQL Hive
Warehouse location
        .appName("SparkSessionExample") // name of the Spark application
        .getOrCreate() // get the existing session or create a new one
```

The preceding builder will try to get an existing SparkSession or create a new one. Then the newly created SparkSession will be assigned as the global default.

 By the way, when using spark-shell, you don't need to create a SparkSession explicitly, because it's already created and accessible with the spark variable.

Creating a DataFrame is probably the most important task in every data analytics task. Spark provides a read() method that can be used to read data from numerous sources in various formats such as CSV, JSON, Avro, and JDBC. For example, the following code snippet shows how to read a CSV file and create a Spark DataFrame:

```
val dataDF = spark.read
        .option("header", "true") // we read the header to know the column
and structure
        .option("inferSchema", "true") // we infer the schema preserved in
the CSV
```

```
       .format("com.databricks.spark.csv") // we're using the CSV reader
from DataBricks
       .load("data/inputData.csv") // Path of the CSV file
       .cache // [Optional] cache if necessary
```

Once a DataFrame is created, we can see a few samples (that is, rows) by invoking
the show() method, as well as print the schema using the printSchema() method.
Invoking describe().show() will show the statistics about the DataFrame:

```
dataDF.show() // show first 10 rows
dataDF.printSchema() // shows the schema (including column name and type)
dataDF.describe().show() // shows descriptive statistics
```

In many cases, we have to use the spark.implicits._ package, which is one of the most
useful imports. It is handy, with a lot of implicit methods for converting Scala objects to
datasets and vice versa. Once we have created a DataFrame, we can create a view
(temporary or global) for performing SQL using either
the ceateOrReplaceTempView() method or the createGlobalTempView() method,
respectively:

```
dataDF.createOrReplaceTempView("myTempDataFrame") // create or replace a
local temporary view with dataDF
dataDF.createGlobalTempView("myGloDataFrame") // create a global temporary
view with dataframe dataDF
```

Now a SQL query can be issued to see the data in tabular format:

```
spark.sql("SELECT * FROM myTempDataFrame")// will show all the records
```

To drop these views, spark.catalog.dropTempView("myTempDataFrame") or
spark.catalog.dropGlobalTempView("myGloDataFrame"), respectively, can be
invoked. By the way, once you're done simply invoking the spark.stop() method, it will
destroy the SparkSession and all the resources allocated by the Spark application.
Interested readers can read detailed API documentation at https://spark.apache.org/ to
get more information.

Reading the training dataset

There is a `Cryotherapy.xlsx` Excel file, which contains data as well as data usage agreement texts. So, I just copied the data and saved it in a CSV file named `Cryotherapy.csv`. Let's start by creating `SparkSession`—the gateway to access Spark:

```
val spark = SparkSession
       .builder
       .master("local[*]")
       .config("spark.sql.warehouse.dir", "/temp")
       .appName("CryotherapyPrediction")
       .getOrCreate()

import spark.implicits._
```

Then let's read the training set and see a glimpse of it:

```
var CryotherapyDF = spark.read.option("header", "true")
              .option("inferSchema", "true")
              .csv("data/Cryotherapy.csv")
```

Let's take a look to see if the preceding CSV reader managed to read the data properly, including header and types:

```
CryotherapyDF.printSchema()
```

As seen from the following screenshot, the schema of the Spark DataFrame has been correctly identified. Also, as expected, all the features of my ML algorithms are numeric (in other words, in integer or double format):

```
root
 |-- sex: integer (nullable = true)
 |-- age: integer (nullable = true)
 |-- Time: double (nullable = true)
 |-- Number_of_Warts: integer (nullable = true)
 |-- Type: integer (nullable = true)
 |-- Area: integer (nullable = true)
 |-- Result_of_Treatment: integer (nullable = true)
```

A snapshot of the dataset can be seen using the `show()` method. We can limit the number of rows; here, let's say 5:

```
CryotherapyDF.show(5)
```

The output of the preceding line of code shows the first five samples of the DataFrame:

```
+---+---+-----+---------------+----+----+-----------------+
|sex|age| Time|Number_of_Warts|Type|Area|Result_of_Treatment|
+---+---+-----+---------------+----+----+-----------------+
|  1| 35| 12.0|              5|   1| 100|                0|
|  1| 29|  7.0|              5|   1|  96|                1|
|  1| 50|  8.0|              1|   3| 132|                0|
|  1| 32|11.75|              7|   3| 750|                0|
|  1| 67| 9.25|              1|   1|  42|                0|
+---+---+-----+---------------+----+----+-----------------+
only showing top 5 rows
```

Preprocessing and feature engineering

As per the dataset description on the UCI machine learning repository, there are no null values. Also, the Spark ML-based classifiers expect numeric values to model them. The good thing is that, as seen in the schema, all the required fields are numeric (that is, either integers or floating point values). Also, the Spark ML algorithms expect a label column, which in our case is Result_of_Treatment. Let's rename it to label using the Spark-provided withColumnRenamed() method:

```
//Spark ML algorithm expect a 'label' column, which is in our case
'Survived". Let's rename it to 'label'
CryotherapyDF = CryotherapyDF.withColumnRenamed("Result_of_Treatment",
"label")
CryotherapyDF.printSchema()
```

All the Spark ML-based classifiers expect training data containing two objects called label (which we already have) and features. We have seen that we have six features. However, those features have to be assembled to create a feature vector. This can be done using the VectorAssembler() method. It is one kind of transformer from the Spark ML library. But first we need to select all the columns except the label column:

```
val selectedCols = Array("sex", "age", "Time", "Number_of_Warts", "Type",
"Area")
```

Then we instantiate a `VectorAssembler()` transformer and transform as follows:

```scala
val vectorAssembler = new VectorAssembler()
        .setInputCols(selectedCols)
        .setOutputCol("features")
val numericDF = vectorAssembler.transform(CryotherapyDF)
                .select("label", "features")
numericDF.show()
```

As expected, the last line of the preceding code segment shows the assembled DataFrame having `label` and `features`, which are needed to train an ML algorithm:

```
+-----+--------------------+
|label|            features|
+-----+--------------------+
|    0|[1.0,35.0,12.0,5....|
|    1|[1.0,29.0,7.0,5.0...|
|    0|[1.0,50.0,8.0,1.0...|
|    0|[1.0,32.0,11.75,7...|
|    0|[1.0,67.0,9.25,1....|
|    1|[1.0,41.0,8.0,2.0...|
|    0|[1.0,36.0,11.0,2....|
|    0|[1.0,59.0,3.5,3.0...|
|    1|[1.0,20.0,4.5,12....|
|    0|[2.0,34.0,11.25,3...|
+-----+--------------------+
only showing top 10 rows
```

Preparing training data and training a classifier

Next, we separate the training set and test sets. Let's say that 80% of the training set will be used for the training and the other 20% will be used to evaluate the trained model:

```scala
val splits = numericDF.randomSplit(Array(0.8, 0.2))
val trainDF = splits(0)
val testDF = splits(1)
```

Instantiate a decision tree classifier by specifying impurity, max bins, and the max depth of the trees. Additionally, we set the `label` and `feature` columns:

```
val dt = new DecisionTreeClassifier()
        .setImpurity("gini")
        .setMaxBins(10)
        .setMaxDepth(30)
        .setLabelCol("label")
        .setFeaturesCol("features")
```

Now that the data and the classifier are ready, we can perform the training:

```
val dtModel = dt.fit(trainDF)
```

Evaluating the model

Since it's a binary classification problem, we need the `BinaryClassificationEvaluator()` estimator to evaluate the model's performance on the test set:

```
val evaluator = new BinaryClassificationEvaluator()
        .setLabelCol("label")
```

Now that the training is completed and we have a trained decision tree model, we can evaluate the trained model on the test set:

```
val predictionDF = dtModel.transform(testDF)
```

Finally, we compute the classification accuracy:

```
val accuracy = evaluator.evaluate(predictionDF)
println("Accuracy =  " + accuracy)
```

You should experience about 96% classification accuracy:

```
Accuracy =  0.9675436785432
```

Finally, we stop the `SparkSession` by invoking the `stop()` method:

```
spark.stop()
```

We have managed to achieve about 96% accuracy with minimum effort. However, there are other performance metrics such as precision, recall, and F1 measure. We will discuss them in upcoming chapters. Also, if you're a newbie to ML and haven't understood all the steps in this example, don't worry. We'll recap all of these steps in other chapters with various other examples.

Summary

In this chapter, we have learned some basic concepts of ML, which is used to solve a real-life problem. We started with a brief introduction to ML including a basic learning workflow, the ML rule of thumb, and different learning tasks, and then we gradually covered important ML tasks such as supervised learning, unsupervised learning, and reinforcement learning. Additionally, we discussed Scala-based ML libraries. Finally, we have seen how to get started with machine learning with Scala and Spark ML by solving a simple classification problem.

Now that we know basic ML and Scala-based ML libraries, we can start learning in a more structured way. In the next chapter, we will learn about regression analysis techniques. Then we will develop a predictive analytics application for predicting slowness in traffic using linear regression and generalized linear regression algorithms.

Scala for Regression Analysis 2

In this chapter, we will learn regression analysis in detail. We will start learning from the regression analysis workflow followed by the **linear regression** (**LR**) and **generalized linear regression** (**GLR**) algorithms. Then we will develop a regression model for predicting slowness in traffic using LR and GLR algorithms and their Spark ML-based implementation in Scala. Finally, we will learn the hyperparameter tuning with cross-validation and the grid searching techniques. Concisely, we will learn the following topics throughout this end-to-end project:

- An overview of regression analysis
- Regression analysis algorithms
- Learning regression analysis through examples
- Linear regression
- Generalized linear regression
- Hyperparameter tuning and cross-validation

Technical requirements

Make sure Scala 2.11.x and Java 1.8.x are installed and configured on your machine.

The code files of this chapters can be found on GitHub:

```
https://github.com/PacktPublishing/Machine-Learning-with-Scala-Quick-Start-
Guide/tree/master/Chapter02
```

Check out the following video to see the Code in Action:

```
http://bit.ly/2GLlQTl
```

An overview of regression analysis

In the previous chapter, we already gained some basic understanding of the **machine learning (ML)** process, as we have seen the basic distinction between regression and classification. Regression analysis is a set of statistical processes for estimating the relationships between a set of variables called a dependent variable and one or multiple independent variables. The values of dependent variables depend on the values of independent variables.

A regression analysis technique helps us to understand this dependency, that is, how the value of the dependent variable changes when any one of the independent variables is changed, while the other independent variables are held fixed. For example, let's assume that there will be more savings in someone's bank when they grow older. Here, the amount of **Savings** (say in million $) depends on age (that is, **Age** in years, for example):

Age (years)	Savings (million $)
40	1.5
50	5.5
60	10.8
70	6.7

So, we can plot these two values in a 2D plot, where the dependent variable (**Savings**) is plotted on the *y*-axis and the independent variable (**Age**) should be plotted on the *x*-axis. Once these data points are plotted, we can see correlations. If the theoretical chart indeed represents the impact of getting older on savings, then we'll be able to say that the older someone gets, the more savings there will be in their bank account.

Now the question is how can we tell the degree to which age helps someone to get more money in their bank account? To answer this question, one can draw a line through the middle of all of the data points on the chart. This line is called the regression line, which can be calculated precisely using a regression analysis algorithm. A regression analysis algorithm takes either discrete or continuous (or both) input features and produces continuous values.

> A classification task is used for predicting the label of the class attribute, while a regression task is used for making a numeric prediction of the class attribute.

Making a prediction using such a regression model on unseen and new observations is like creating a data pipeline with multiple components working together, where we observe an algorithm's performance in two stages: learning and inference. In the whole process and for making the predictive model a successful one, data acts as the first-class citizen in all ML tasks.

Learning

One of the important task at the learning stage is to prepare and convert the data into feature vectors (vectors of numbers out of each feature). Training data in feature vector format can be fed into the learning algorithms to train the model, which can be used for inferencing. Typically, and of course based on data size, running an algorithm may take hours (or even days) so that the features converge into a useful model as shown in the following diagram:

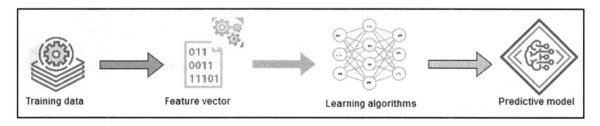

Learning and training a predictive model—it shows how to generate the feature vectors from the training data to train the learning algorithm that produces a predictive model

Inferencing

In the inference stage, the trained model is used for making intelligent use of the model, such as predicting from never-before-seen data, making recommendations, and deducing future rules. Typically, it takes less time compared to the learning stage and sometimes even in real time. Thus, inferencing is all about testing the model against new (that is, unobserved) data and evaluating the performance of the model itself, as shown in the following diagram:

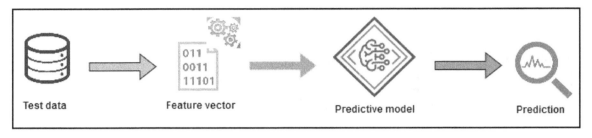

Inferencing from an existing model towards predictive analytics (feature vectors are generated from unknown data for making predictions)

In summary, when using regression analysis the goal is to predict a continuous target variable. Now that we know how to construct a basic workflow for a supervised learning task, knowing a little about available regression algorithms will provide a bit more concrete information on how to apply these regression algorithms.

Regression analysis algorithms

There are numerous algorithms proposed and available, which can be used for the regression analysis. For example, LR tries to find relationships and dependencies between variables. It models the relationship between a continuous dependent variable y (that is, a label or target) and one or more independent variables, x, using a linear function. Examples of regression algorithms include the following:

- **Linear regression (LR)**
- **Generalized linear regression (GLR)**
- **Survival regression (SR)**
- **Isotonic regression (IR)**
- **Decision tree regressor (DTR)**
- **Random forest regression (RFR)**
- **Gradient boosted trees regression (GBTR)**

We start by explaining regression with the simplest LR algorithm, which models the relationship between a dependent variable, *y*, which involves a linear combination of interdependent variables, *x*:

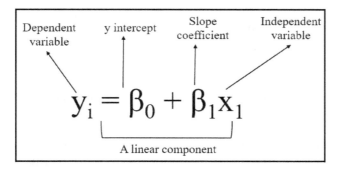

In the preceding equation letters, β_0 and β_1 are two constants for *y*-axis intercept and the slope of the line, respectively. LR is about learning a model, which is a linear combination of features of the input example (data points).

Take a look at the following graph and imagine that the red line is not there. We have a few dotted blue points (data points). Can we reasonably develop a machine learning (regression) model to separate most of them? Now, if we draw a straight line between two classes of data, those get almost separated, don't they? Such a line (red in our case) is called the decision boundary, which is also called the regression line in the case of regression analysis (see the following example for more):

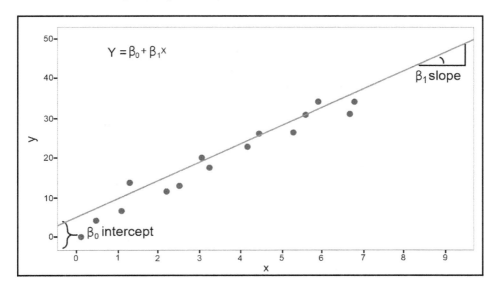

If we are given a collection of labeled examples, say $\{(X_i, y_i)\}_{i=1}^{N}$, where N is the number of samples in the dataset, x_i is the D-dimensional feature vector of the samples $i = 1, 2 \dots N$, and y_i is a real-valued $y \in R$, where R denotes the set of all real numbers called the target variable and every feature x_i is a real number. Then combining these, the next step is to build the following mathematical model, f:

$$f_{w,b}(x) = wx + b$$

Here, w is a D-dimensional parameterized vector and b is a real number. The notation $f_{w,b}$ signifies that the model f is parameterized by values w and b. Once we have a well-defined model, it can now be used for making a prediction of unknown y for a given x, that is, $y \leftarrow f_{w,b}(x)$. However, there is an issue, as since the model is parametrized with two different values (w, b), this will mean the model tends to produce two different predictions when applied to the same sample, even when coming from the same distribution.

Literally, it can be referred as an optimization problem—where the objective is to find the optimal (that is, minimum, for example) values $(\hat{w}, \hat{b})$ such that the optimal values of parameters will mean the model tends to make more accurate predictions. In short, in the LR model, we intend to find the optimal values for $\hat{w}$ and $\hat{b}$ to minimize the following objective function:

$$\min_{w,b} \frac{1}{N} \sum_{i=1,2,3 \dots N} (f_{w,b}(X_i) - y_i)^2$$

In the preceding equation, the expression $(f_{w,b}(X_i) - y_i)^2$ is called the **loss function**, which is a measure of penalty (that is, error or loss) for giving the wrong prediction for sample i. This loss function is in the form of squared error loss. However, other loss functions can be used too, as outlined in the following equations:

$$SE = \sum_{i=1}^{N} (y_i - F(x_i))^2 \dots \dots (1)$$

$$AE = \sum_{i=1}^{N} |y_i - F(x_i)| \dots \dots (2)$$

The **squared error** (**SE**) in equation 1 is called L_2 loss, which is the default loss function for the regression analysis task. On the other hand, the **absolute error** (**AE**) in equation (2) is called L_1 loss.

> In cases where the dataset has many outliers, using L_1 loss is recommend more than L_2, because L_1 is more robust against outliers.

All model-based learning algorithms have a loss function associated with them. Then we try to find the best model by minimizing the cost function. In our LR case, the cost function is defined by the average loss (also called empirical risk), which can be formulated as the average of all penalties obtained by fitting the model to the training data, which may contain many samples.

Figure 4 shows an example of simple linear regression. Let's say the idea is to predict the amount of **Savings** versus **Age**. So, in this case, we have one independent variable x (that is, a set of 1D data points and, in our case, the **Age**) and one dependent variable, y (amount of **Savings (in millions $)**). Once we have a trained regression model, we can use this line to predict the value of the target y_l for a new unlabeled input example, x_l. However, in the case of D -dimensional feature vectors (for example, *2D* or *3D*), it would be a plane (for *2D*) or a hyperplane (for *>=3D*):

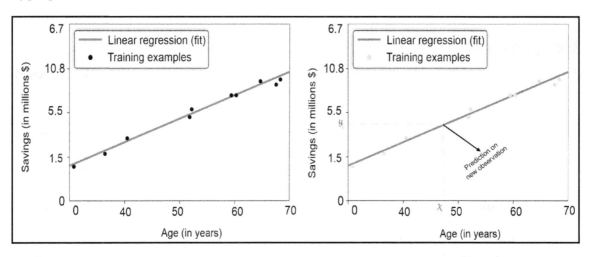

Figure 4: A regression line separates data points to solve Age versus the amount of Savings: i) the left model separates data points based on training data: ii) the right model predicts for an unknown observation

Now you see why it is important to have the requirement that the regression hyperplane lies as close to the training examples as possible: if the blue line in *Figure 4* (the model on the right) is far away from the blue dots, the prediction y_l is less likely to be correct. The best fit line, which is expected to pass through most of the data points, is the result of the regression analysis. However, in practice it does not pass through all of the data points because of the existence of regression errors.

Regression error is the distance between any data points (actual) and the line (predicted).

Since solving a regression problem is itself an optimization problem, we expect a smaller margin for errors as possible because smaller errors contribute towards higher predictive accuracy, while predicting unseen observations. Although an LR algorithm is not so efficient in many cases, the nicest thing is that an LR model usually does not overfit, which is unlikely for a more complex model.

In the previous chapter, we discussed overfitting (a phenomenon whereby a model that shows a model predicts very well during the training but makes more errors when applied to test set) and underfitting (if your training error is low and your validation error is high, then your model is most likely overfitting your training data). Often these two phenomena occur due to bias and variance.

Performance metrics

To measure the predictive performance of a regression model, several metrics are proposed and in use in terms of regression errors, which can be outlined as follows:

- **Mean squared error (MSE)**: It is the measure of the difference between the predicted and estimated values, that is, how close a fitted line is to data points. The smaller the MSE, the closer the fit is to the data.
- **Root mean squared error (RMSE)**: It is the square root of the MSE but has the same units as the quantity plotted on the vertical axis.
- **R-squared**: It is the coefficient of determination for assessing how close the data is to the fitted regression line ranges between 0 and 1. The higher the R-squared, the better the model fits your data.
- **Mean absolute error (MAE)**: It is a measure of *accuracy* for continuous variables without considering their direction. The smaller the MAE, the better the model fits your data.

Now that we know how a regression algorithm works and how to evaluate the performance using several metrics, the next important task is to apply this knowledge to solve a real-life problem.

Learning regression analysis through examples

In the previous section, we discussed a simple real-life problem (that is, **Age** versus **Savings**). However, in practice, there are several real-life problems where more factors and parameters (that is, data properties) are involved, where regression can be applied too. Let's first introduce a real-life problem. Imagine that you live in Sao Paulo, a city in Brazil, where every day several hours of your valuable time are wasted because of unavoidable reasons such as an immobilized bus, broken truck, vehicle excess, accident victim, overtaking, fire vehicles, incident involving dangerous freight, lack of electricity, fire, and flooding.

Now, to measure how many man hours get wasted, we can we develop an automated technique, which will predict the slowness of traffic such that you can avoid certain routes or at least get some rough estimation of how long it'll take you to reach some point in the city. A predictive analytics application using machine learning is probably one of the preferred solutions for predicting such slowness. Yes, for that we'll use the behavior of the urban traffic of the city of Sao Paulo in Brazil dataset in the next section.

Description of the dataset

The dataset is downloaded from `https://archive.ics.uci.edu/ml/datasets/ Behavior+of+the+urban+traffic+of+the+city+of+Sao+Paulo+in+Brazil`. It contains the records of behavior of the urban traffic of the city of Sao Paulo in Brazil between December 14, 2009 and December 18, 2009. The dataset has the following features:

- **Hour**: Total hours spent on the road
- **Immobilized bus**: Number of immobilized buses
- **Broken truck**: Number of broken trucks
- **Vehicle excess**: Number of redundant vehicles
- **Accident victim**: Number of accident victims on the road or road side
- **Running over**: Number of running over or taking over cases
- **Fire vehicles**: Number of fire trucks and vehicles

- **Occurrence involving freight**: Number of goods transported in bulk by trucks
- **Incident involving dangerous freight**: Number of transporter bulk trucks involved in accident
- **Lack of electricity**: Number of hours without electricity in the affected areas
- **Fire**: Number of fire incidents
- **Point of flooding**: Number of points of flooding areas
- **Manifestations**: Number of places showing construction work ongoing or dangerous signs
- **Defect in the network of trolleybuses**: Number of defects in the network of trolley buses
- **Tree on the road**: Number of trees on the road or road side that create obstacles
- **Semaphore off**: Number of mechanical gadgets with arms, lights, or flags that are used as a signal
- **Intermittent semaphore**: Number of mechanical gadgets with arms, lights, or flags that are used as a signal for a specific period of time
- **Slowness in traffic**: Number of average hours people got stuck in traffic because of the preceding reasons

The last feature is the target column, which we want to predict. Since I used this dataset, I would like to acknowledge the following publication:

Ferreira, R. P., Affonso, C., & Sassi, R. J. (2011, November). Combination of Artificial Intelligence Techniques for Prediction the Behavior of Urban Vehicular Traffic in the City of Sao Paulo. In 10th Brazilian Congress on Computational Intelligence (CBIC) - Fortaleza, Brazil. (pp.1-7), 2011.

Exploratory analysis of the dataset

First, we read the training set for the **exploratory data analysis (EDA)**. Readers can refer to the `EDA.scala` file for this. Once extracted, there will be a CSV file named `Behavior of the urban traffic of the city of Sao Paulo in Brazil.csv`. Let's rename the file as `UrbanTraffic.csv`. Also, `Slowness in traffic (%)`, which is the last column, represents the percentage of slowness in an unusual format: it represents the real number with a comma (`,`), for example, `4,1` instead of `4.1`. So I replaced all instances of a comma (`,`) in that column with a period (`.`). Otherwise, the Spark CSV reader will treat the column as a `String` type:

```scala
val filePath= "data/UrbanTraffic.csv"
```

First, let's load, parse, and create a DataFrame using the `read.csv()` method but with the Databricks CSV format (also known as `com.databricks.spark.csv`) by setting it to read the header of the CSV file, which is directly applied to the columns' names of the DataFrame created; and the `inferSchema` property is set to `true`, because if you don't specify the `inferSchema` configuration explicitly, the float values would be treated as strings. This might cause `VectorAssembler` to raise an exception such as `java.lang.IllegalArgumentException: Data type StringType is not supported`:

```
val rawTrafficDF = spark.read
    .option("header", "true")
    .option("inferSchema", "true")
    .option("delimiter", ";")
    .format("com.databricks.spark.csv")
    .load("data/UrbanTraffic.csv")
    .cache
```

Now let's print the schema of the DataFrame we just created to check to make sure the structure is preserved:

```
rawTrafficDF.printSchema()
```

As seen from the following screenshot, the schema of the Spark DataFrame has been correctly identified. Also, as expected, all the features of my ML algorithms are numeric (in other words, in integer or double format):

```
root
 |-- Hour (Coded): integer (nullable = true)
 |-- Immobilized bus: integer (nullable = true)
 |-- Broken Truck: integer (nullable = true)
 |-- Vehicle excess: integer (nullable = true)
 |-- Accident victim: integer (nullable = true)
 |-- Running over: integer (nullable = true)
 |-- Fire vehicles: integer (nullable = true)
 |-- Occurrence involving freight: integer (nullable = true)
 |-- Incident involving dangerous freight: integer (nullable = true)
 |-- Lack of electricity: integer (nullable = true)
 |-- Fire: integer (nullable = true)
 |-- Point of flooding: integer (nullable = true)
 |-- Manifestations: integer (nullable = true)
 |-- Defect in the network of trolleybuses: integer (nullable = true)
 |-- Tree on the road: integer (nullable = true)
 |-- Semaphore off: integer (nullable = true)
 |-- Intermittent Semaphore: integer (nullable = true)
 |-- Slowness in traffic (%): double (nullable = true)
```

You can see that none of the columns are categorical features. So, we don't need any numeric transformation. Now let's see how many rows there are in the dataset using the `count()` method:

```
println(rawTrafficDF.count())
```

This gives a 135 sample count. Now let's see a snapshot of the dataset using the `show()` method, but with only some selected columns so that it can make more sense rather than showing all of them. But feel free to use `rawTrafficDF.show()` to see all columns:

```
rawTrafficDF.select("Hour (Coded)", "Immobilized bus", "Broken Truck",
                    "Vehicle excess", "Fire", "Slowness in traffic
(%)").show(5)
```

As the `Slowness in traffic (%)` column contains continuous values, we have to deal with a regression task. Now that we have seen a snapshot of the dataset, it would be worth seeing some other statistics such as average claim or loss, minimum, and maximum loss of Spark SQL using the `sql()` interface:

```
+------------+---------------+------------+--------------+----+---------------------+
|Hour (Coded)|Immobilized bus|Broken Truck|Vehicle excess|Fire|Slowness in traffic (%)|
+------------+---------------+------------+--------------+----+---------------------+
|           1|              0|           0|             0|   0|                  4.1|
|           2|              0|           0|             0|   0|                  6.6|
|           3|              0|           0|             0|   0|                  8.7|
|           4|              0|           0|             0|   0|                  9.2|
|           5|              0|           0|             0|   0|                 11.1|
+------------+---------------+------------+--------------+----+---------------------+
only showing top 5 rows
```

However, before that, let's rename the last column from `Slowness in traffic (%)` to `label`, since the ML model will complain about it. Even after using `setLabelCol` on the regression model, it still looks for a column called `label`. This introduces a disgusting error saying `org.apache.spark.sql.AnalysisException: cannot resolve 'label' given input columns`:

```
var newTrafficDF = rawTrafficDF.withColumnRenamed("Slowness in traffic
(%)", "label")
```

Since we want to execute some SQL query, we need to create a temporary view so that the operation can be performed in-memory:

```
newTrafficDF.createOrReplaceTempView("slDF")
```

Now let's average the slowness in the form of a percentage (the deviation with standard hours):

```
spark.sql("SELECT avg(label) as avgSlowness FROM slDF").show()
```

The preceding line of code should show a 10% delay on average every day across routes and based on other factors:

```
+------------------+
|    avgSlowness   |
+------------------+
|10.051851851851858|
+------------------+
```

Also, we can see the number of flood points in the city. However, for that we might need some extra effort by changing the column name into a single string since it's a multi-string containing spaces, so SQL won't be able to resolve it:

```
newTrafficDF = newTrafficDF.withColumnRenamed("Point of flooding",
"NoOfFloodPoint")
spark.sql("SELECT max(NoOfFloodPoint) FROM slDF").show()
```

This should show as many as seven flood points that could be very dangerous:

```
+------------------+
|max(NoOfFloodPoint)|
+------------------+
|                 7|
+------------------+
```

However, the `describe()` method will give these types of statistics more flexibly. Let's do it for the selected columns:

```
rawTrafficDF.select("Hour (Coded)", "Immobilized bus", "Broken Truck",
                    "Point of flooding", "Fire", "Slowness in traffic (%)")
                    .describe().show()
```

So, we can see that the slowness varies between 3.4 and 23.4, which is quite high. This is why we need efficient data processing steps so that such a relation can be preserved. Now let's focus on the data preprocessing instead:

```
+-------+-----------------+-------------------+------------------+------------------+------------------+------------------------+
|summary|     Hour (Coded)|     Immobilized bus|      Broken Truck|  Point of flooding|              Fire|Slowness in traffic (%)|
+-------+-----------------+-------------------+------------------+------------------+------------------+------------------------+
|  count|              135|                135|               135|               135|               135|                     135|
|   mean|             14.0|0.34074074074074073|0.8740740740740741|0.11851851851851852|0.007407407407407408|      10.051851851851858|
| stddev|7.817889931757458|0.6597485156670081|1.1024371212964403|0.712907143184692|0.08606629658238704|       4.363242869254161|
|    min|                1|                  0|                 0|                 0|                 0|                     3.4|
|    max|               27|                  4|                 5|                 7|                 1|                    23.4|
+-------+-----------------+-------------------+------------------+------------------+------------------+------------------------+
```

Feature engineering and data preparation

Now that we have seen some properties of the dataset and since there're no null values or categorical features, we don't need any other preprocessing or intermediate transformations. We just need to do some feature engineering before we can have our training and test sets.

The first step before getting these sets is to prepare training data that is consumable by the Spark regression model. For this, Spark classification and regression algorithms expect two components called `features` and `label`. Fortunately, we already have the `label` column. Next, the `features` column has to contain the data from all the columns except the `label` column, which can be achieved using the `VectorAssembler()` transformer.

Since all the columns are numeric, we can use `VectorAssembler()` directly from the Spark ML library to transform a given list of columns into a single vector column. So, let's collect the list of desirable columns. As you may have guessed, we'll have to exclude the `label` column, which can be done using the `dropRight()` method of standard Scala:

```scala
val colNames = newTrafficDF.columns.dropRight(1)

val assembler = new VectorAssembler()
    .setInputCols(colNames)
    .setOutputCol("features")
```

Now that we have the `VectorAssembler()` estimator, we now call the `transform()` method, which will embed selected columns into a single vector column:

```scala
val assembleDF = assembler.transform(newTrafficDF).select("features",
"label")
assembleDF.show()
```

As expected, the last line of the preceding code segment shows the assembled DataFrame having `label` and `features`, which are needed to train an ML algorithm:

```
+---------------+-----+
|       features|label|
+---------------+-----+
| (17,[0],[1.0])|  4.1|
| (17,[0],[2.0])|  6.6|
| (17,[0],[3.0])|  8.7|
| (17,[0],[4.0])|  9.2|
| (17,[0],[5.0])| 11.1|
| (17,[0],[6.0])| 10.9|
| (17,[0],[7.0])|  8.3|
| (17,[0],[8.0])|  8.2|
| (17,[0],[9.0])|  7.6|
|(17,[0],[10.0])|  6.4|
+---------------+-----+
only showing top 10 rows
```

We can now proceed to generate separate training and test sets. Additionally, we can cache both the sets for faster in-memory access. We use 60% of the data to train the model and the other 40% will be used to evaluate the model:

```
val seed = 12345L
val splits = data.randomSplit(Array(0.60, 0.40), seed)
val (trainingData, validationData) = (splits(0), splits(1))

trainingData.cache // cache in memory for quicker access
validationData.cache // cache in memory for quicker access
```

That is all we need before we start training the regression models. At first, we start training the LR model and evaluate the performance.

Linear regression

In this section, we will develop a predictive analytics model for predicting slowness in traffic for each row of the data using an LR algorithm. First, we create an LR estimator as follows:

```
val lr = new LinearRegression()
        .setFeaturesCol("features")
        .setLabelCol("label")
```

Then we invoke the `fit()` method to perform the training on the training set as follows:

```
println("Building ML regression model")
val lrModel = lr.fit(trainingData)
```

Now we have the fitted model, which means it is now capable of making predictions. So, let's start evaluating the model on the training and validation sets and calculating the RMSE, MSE, MAE, R squared, and so on:

```
println("Evaluating the model on the test set and calculating the
regression metrics")
// ********************************************************************
val trainPredictionsAndLabels = lrModel.transform(testData).select("label",
"prediction")
                                            .map {case Row(label: Double,
prediction: Double)
                                            => (label, prediction)}.rdd

val testRegressionMetrics = new
RegressionMetrics(trainPredictionsAndLabels)
```

Great! We have managed to compute the raw prediction on the training and the test sets. Now that we have both the performance metrics on both training and validation sets, let's observe the results of the training and the validation sets:

```
val results =
"\n=====================================================================\n"
+
        s"TrainingData count: ${trainingData.count}\n" +
        s"TestData count: ${testData.count}\n" +
"=====================================================================\n" +
        s"TestData MSE = ${testRegressionMetrics.meanSquaredError}\n" +
        s"TestData RMSE = ${testRegressionMetrics.rootMeanSquaredError}\n" +
        s"TestData R-squared = ${testRegressionMetrics.r2}\n" +
        s"TestData MAE = ${testRegressionMetrics.meanAbsoluteError}\n" +
        s"TestData explained variance =
${testRegressionMetrics.explainedVariance}\n" +
"=====================================================================\n"
println(results)
```

The preceding code segment should show something similar. Although, because of the randomness, you might experience slightly different output:

```
=====================================================================
 TrainingData count: 80
 TestData count: 55
=====================================================================
 TestData MSE = 7.904822843038552
```

```
TestData RMSE = 2.8115516788845536
TestData R-squared = 0.3699441827613118
TestData MAE = 2.2173672546414536
TestData explained variance = 20.293395978801147
=====================================================================
```

Now that we have the prediction on the test set as well, however, we can't directly say if it's a good or optimal regression model. To improve the result further with lower MAE, Spark also provides the generalized version of linear regression implementation called GLR.

Generalized linear regression (GLR)

In an LR, the output is assumed to follow a Gaussian distribution. In contrast, in **generalized linear models** (**GLMs**), the response variable Y_i follows some random distribution from a parametric set of probability distributions of a certain form. As we have seen in the previous example, following and creating a GLR estimator will not be difficult:

```
val glr = new GeneralizedLinearRegression()
    .setFamily("gaussian")//continuous value prediction (or gamma)
    .setLink("identity")//continuous value prediction (or inverse)
    .setFeaturesCol("features")
    .setLabelCol("label")
```

For the GLR-based prediction, the following response and identity link functions are supported based on data types (source: `https://spark.apache.org/docs/latest/ml-classification-regression.html#generalized-linear-regression`):

Family	Response Type	Supported Links
Gaussian	Continuous	Identity*, Log, Inverse
Binomial	Binary	Logit*, Probit, CLogLog
Poisson	Count	Log*, Identity, Sqrt
Gamma	Continuous	Inverse*, Idenity, Log
Tweedie	Zero-inflated continuous	Power link function
* Canonical Link		

Then we invoke the `fit()` method to perform the training on the training set as follows:

```
println("Building ML regression model")
val glrModel = glr.fit(trainingData)
```

The current implementation through the `GeneralizedLinearRegression` interface in Spark supports up to 4,096 features only. Now that we have the fitted model (which means it is now capable of making predictions), let's start evaluating the model on training and validation sets and calculating the RMSE, MSE, MAE, R squared, and so on:

```
// **********************************************************************
println("Evaluating the model on the test set and calculating the
regression metrics")
// **********************************************************************
val trainPredictionsAndLabels =
glrModel.transform(testData).select("label", "prediction")
                                        .map { case Row(label: Double,
prediction: Double)
                                     => (label, prediction) }.rdd

val testRegressionMetrics = new
RegressionMetrics(trainPredictionsAndLabels)
```

Great! We have managed to compute the raw prediction on the training and the test sets. Now that we have both the performance metrics on both training and test sets, let's observe the result on the train and the validation sets:

```
val results =
"\n====================================================================\n"
+
     s"TrainingData count: ${trainingData.count}\n" +
     s"TestData count: ${testData.count}\n" +
"====================================================================\n" +
     s"TestData MSE = ${testRegressionMetrics.meanSquaredError}\n" +
     s"TestData RMSE = ${testRegressionMetrics.rootMeanSquaredError}\n" +
     s"TestData R-squared = ${testRegressionMetrics.r2}\n" +
     s"TestData MAE = ${testRegressionMetrics.meanAbsoluteError}\n" +
     s"TestData explained variance =
${testRegressionMetrics.explainedVariance}\n" +
"====================================================================\n"
println(results)
```

The preceding code segment should show similar results. Although, because of the randomness, you might experience slightly different output:

```
====================================================================
TrainingData count: 63
```

```
TestData count: 72
=========================================================================
TestData MSE = 9.799660597570348
TestData RMSE = 3.130440958965741
TestData R-squared = -0.1504361865072692
TestData MAE = 2.5046175463628546
TestData explained variance = 19.241059408685135
=========================================================================
```

Using GLR, we can see a slightly worse MAE value and also the RMSE is higher. If you see these two examples, we have not got to tune the hyperparameters but simply let the models train and evaluate a single value of each parameter. We could even use a regularization parameter for reducing overfitting. However, the performance of an ML pipeline often improves with the hyperparameter tuning, which is usually done with grid search and cross-validation. In the next section, we will discuss how we can get even better performance with the cross-validated models.

Hyperparameter tuning and cross-validation

In machine learning, the term hyperparameter refers to those parameters that cannot be learned from the regular training process directly. These are the various knobs that you can tweak on your machine learning algorithms. Hyperparameters are usually decided by training the model with different combinations of the parameters and deciding which ones work best by testing them. Ultimately, the combination that provides the best model would be our final hyperparameters. Setting hyperparameters can have a significant influence on the performance of the trained models.

On the other hand, cross-validation is often used in conjunction with hyperparameter tuning. Cross-validation (also know as rotation estimation) is a model validation technique for assessing the quality of the statistical analysis and results. Cross-validation helps to describe a dataset to test the model in the training phase using the validation set.

Hyperparameter tuning

Unfortunately, there is no shortcut or straightforward way of choosing the right combination of hyperparameters based on a clear recipe—of course, experience helps. For example, while training a random forest, Matrix factorization, k-means, or a logistic/LR algorithm might be appropriate. Here are some typical examples of such hyperparameters:

- Number of leaves, bins, or depth of a tree in tree-based algorithms
- Number of iterations

- Regularization values
- Number of latent factors in a matrix factorization
- Number of clusters in a k-means clustering and so on

Technically, hyperparameters form an *n*-dimensional space called a param-grid, where *n* is the number of hyperparameters. Every point in this space is one particular hyperparameter configuration, which is a hyperparameter vector.

As discussed in `Chapter 1`, *Introduction to Machine Learning with Scala*, overfitting and underfitting are two problematic phenomena in machine learning. Therefore, sometimes full convergence to a best model parameter set is often not necessary and can be even preferred, because an almost-best-fitting model tends to perform better on new data or settings. In other words, if you care for a best fitting model, you really don't need the best parameter set.

In practice, we cannot explore every point in this space, so the grid search over a subset in that space is commonly used. The following diagram shows some high-level idea:

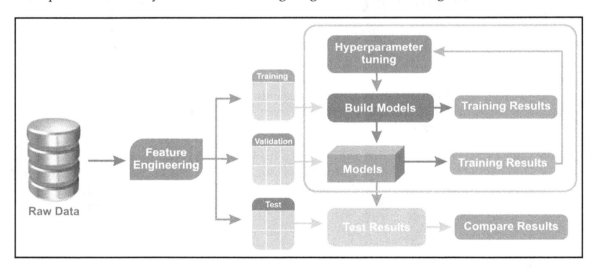

Figure 5: Hyperparameter tuning of ML models

Although there are several approaches for such a scheme, random search or grid search are probably the most well-known techniques used:

- **Grid search**: Using this approach, different hyperparameters are defined in a dictionary that you want to test. Then a param-grid is constructed before feeding them into the ML model such that the training can be performed with the different combinations. Finally, the algorithm tells you for which combination of the hyperparameters you have the highest accuracy.
- **Random search**: As you can understand, training an ML model with all possible combinations of hyperparameters is a very expensive and time consuming operation. However, often we don't have that much flexibility but still we want to tune those parameters. In such a situation, random search could be a workaround. Random search is performed through evaluating *n* uniformly random points in the hyperparameter space, and selecting the right combination for which the model gives the best performance.

Cross-validation

There are two types of cross-validation, called exhaustive cross-validation, which includes leave-p-out cross-validation and leave-one-out cross-validation, and non-exhaustive cross-validation, which is based on K-fold cross-validation and repeated random sub-sampling cross-validation, for example, 5-fold or 10-fold cross-validation, is very common.

In most of the cases, 10-fold cross-validation is used instead of testing on a validation set. Also, the training set should be as large as possible (because more data with quality features are good to train the model) not only to train the model but because about 5 to 10% of the training set can be used for the cross-validation.

Using the K-fold cross-validation technique, the complete training data is split into K subsets. The model is trained on K-1 subsets; hold the last one for the validation. This process is repeated K times so that each time, one of the K subsets is used as the validation set and the other K-1 subsets are used to form the training set. This way, each of the subsets (fold) is used at least once for both training and validation.

Finally, different machine learning models that have been obtained are joined by a bagging (or boosting) scheme for classifiers or by averaging (that is, regression). The following diagram explains the 10-fold cross-validation technique:

Figure 6: 10-fold cross-validation technique

Tuning and cross-validation in Spark ML

In Spark ML, before performing the cross-validation, we need to have a `paramGrid` (that is a grid of parameters). The `ParamGridBuilder` interface is used in order to define the hyperparameter space where `CrossValidator` has to search and finally, `CrossValidator()` takes our pipeline, the hyperparameter space of our LR regressor, and the number of folds for the cross-validation as parameters.

So, let's start creating `paramGrid` by specifying the number of maximum iterations, the value of regularization parameter, the value of tolerance, and the elastic network parameters, as follows for the LR model (since we observed lower MAE for this):

```
// ************************************************************
println("Preparing K-fold Cross Validation and Grid Search")
// ************************************************************
val paramGrid = new ParamGridBuilder()
      .addGrid(lr.maxIter, Array(10, 20, 30, 50, 100, 500, 1000))
      .addGrid(lr.regParam, Array(0.001, 0.01, 0.1))
      .addGrid(lr.tol, Array(0.01, 0.1))
      .build()
```

The regularization parameter reduces overfitting by reducing the variance of your estimated regression parameters. Now, for a better and more stable performance, we can perform 10-fold cross-validation. Since our task is predicting continuous values, we need to define `RegressionEvaluator`, that is, the evaluator for regression, which expects two input columns—`prediction` and `label`—and evaluates the training based on MSE, RMSE, R-squared, and MAE:

```
println("Preparing 10-fold Cross Validation")
val numFolds = 10 //10-fold cross-validation
val cv = new CrossValidator()
      .setEstimator(lr)
      .setEvaluator(new RegressionEvaluator())
      .setEstimatorParamMaps(paramGrid)
      .setNumFolds(numFolds)
```

Fantastic, we have created the cross-validation estimator. Now it's time to train the LR model:

```
println("Training model with the Linear Regression algorithm")
val cvModel = cv.fit(trainingData)
```

By the way, Spark provides a way to save a trained ML model using the `save()` method:

```
// Save the workflow
cvModel.write.overwrite().save("model/LR_model")
```

Then the same model can be restored from the disk using the `load()` method:

```
val sameCVModel = LinearRegressionModel.load("model/LR_model")
```

Then we compute the model's metrics on the test set similar to the LR and GLR models:

```
println("Evaluating the cross validated model on the test set and
calculating the regression metrics")
```

```
val trainPredictionsAndLabelsCV =
cvModel.transform(testData).select("label", "prediction")
                                        .map { case Row(label: Double,
prediction: Double)
                                    => (label, prediction) }.rdd

val testRegressionMetricsCV = new
RegressionMetrics(trainPredictionsAndLabelsCV)
```

Finally, we gather the metrics and print to get some insights:

```
val cvResults =
"\n=======================================================================\n"
+
    s"TrainingData count: ${trainingData.count}\n" +
    s"TestData count: ${testData.count}\n" +
"=======================================================================\n" +
    s"TestData MSE = ${testRegressionMetricsCV.meanSquaredError}\n" +
    s"TestData RMSE = ${testRegressionMetricsCV.rootMeanSquaredError}\n"
+
    s"TestData R-squared = ${testRegressionMetricsCV.r2}\n" +
    s"TestData MAE = ${testRegressionMetricsCV.meanAbsoluteError}\n" +
    s"TestData explained variance =
${testRegressionMetricsCV.explainedVariance}\n" +
"=======================================================================\n"
println(cvResults)
```

The preceding code segment should show something similar. Although, because of the randomness, you might experience slightly different output:

```
=======================================================================
TrainingData count: 80
TestData count: 55
=======================================================================
TestData MSE = 7.889401628365509
TestData RMSE = 2.8088078660466453
TestData R-squared = 0.3510269588724132
TestData MAE = 2.2158433237623667
TestData explained variance = 20.299135214455085
=======================================================================
```

As we can see, both the RMSE and MAE are slightly lower than the non-cross validated LR model. Ideally, we should have experienced even lower values for these metrics. However, due to the small size of the training as well as test sets, probably both the LR and GLR models overfitted. Still, we will try to use robust regression analysis algorithms in Chapter 4, *Scala for Tree-Based Ensemble Techniques*. More specifically, we will try to solve the same problem with decision trees, random forest, and GBTRs.

Summary

In this chapter, we have seen how to develop a regression model for analyzing insurance severity claims using LR and GLR algorithms. We have also seen how to boost the performance of the GLR model using cross-validation and grid search techniques, which give the best combination of hyperparameters. Finally, we have seen some frequently asked questions so that the similar regression techniques can be applied for solving other real-life problems.

In the next chapter, we will see another supervised learning technique called classification through a real-life problem called analyzing outgoing customers through churn prediction. Several classification algorithms will be used for making the prediction in Scala. Churn prediction is essential for businesses as it helps you detect customers who are likely to cancel a subscription, product, or service, which also minimizes customer defection by predicting which customers are likely to cancel a subscription to a service.

3
Scala for Learning Classification

In the previous chapter, we saw how to develop a predictive model for analyzing insurance severity claims as a regression analysis problem. We applied very simple linear regression, as well as **generalized linear regression (GLR)**.

In this chapter, we'll learn about another supervised learning task, called classification. We'll use widely used algorithms such as logistic regression, **Naive Bayes (NB)**, and **Support Vector Machines (SVMs)** to analyze and predict whether a customer is likely to cancel the subscription of their telecommunication contract or not.

In particular, we will cover the following topics:

- Introduction to classification
- Learning classification with a real-life example
- Logistic regression for churn prediction
- SVM for churn prediction
- NB for prediction

Technical requirements

Make sure Scala 2.11.x and Java 1.8.x are installed and configured on your machine.

The code files of this chapters can be found on GitHub:

```
https://github.com/PacktPublishing/Machine-Learning-with-Scala-Quick-Start-
Guide/tree/master/Chapter03
```

Check out the following video to see the Code in Action:
```
http://bit.ly/2ZKVrxH
```

Overview of classification

As a supervised learning task, classification is the problem of identifying which set of observations (sample) belongs to what based on one or more independent variables. This learning process is based on a training set containing observations (or instances) about the class or label of membership. Typically, classification problems are when we are training a model to predict quantitative (but discrete) targets, such as spam detection, churn prediction, sentiment analysis, cancer type prediction, and so on.

Suppose we want to develop a predictive model, which will predict whether a student is competent enough to get admission into computer science based on his/her competency in TOEFL and GRE. Also, suppose we have some historical data in the following range/format:

- **TOEFL**: Between 0 and 100
- **GRE**: Between 0 and 100
- **Admission**: 1 for admitted, 0 if not admitted

Now, to understand whether we can use such simple data to make predictions, let's create a scatter plot by putting all the records with **Admitted** and **Rejected** as the dependent variables and **TOEFL** and **GRE** as the independent variables, like so:

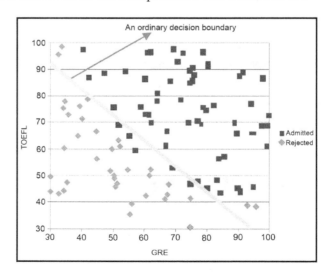

By looking at the data points (imagine that the diagonal line in the graph is not there), we can reasonably develop a linear model to separate most of the data points. Now, if we draw a straight line between two classes of data, those almost get separated. Such a line (green, in our case) is called the decision boundary. So, if the decision boundary has reasonably separate maximal data points, it can be used for making predictions on unseen data, and we can also say that the data point above the line we predicted is competent for admission, and below the line we predict that the students are not competent enough.

Although this example is for a basic head-start into regression analysis, separating missions of data points is not very easy. Thus, to calculate where to draw the line for separating such a huge number of data points, we can use logistic regression or other classification algorithms that we will discuss in upcoming sections. We'll also see that drawing an ordinary straight line might not be the right one, and therefore we often have to draw curved lines.

If we look at the admission-related data plot carefully, maybe a straight line is not the best way of separating each data point—a curved line would be better, as shown in the following graph:

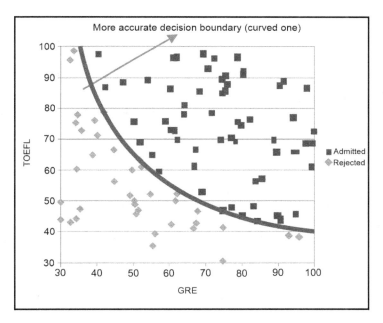

However, to get a curved decision boundary, we have to change not only the function (called the decision boundary function) that's responsible from being linear to some high-order polynomial, but also the data to be a second-degree polynomial.

This means that we have to model our problem as a logistic regression model. That is, we need to change the data from *{GRE, TOEFL}* format to a quadratic function format, *{GRE, GRE^2, TOEFL, TOEFL^2, GRE∗TOEFL}*. However, doing so in a hand-crafted way is cumbersome and will not be possible for large datasets. Fortunately, Spark MLlib has numerous algorithms implemented for modeling such problems and for solving other classification problems, including the following:

- **Logistic regression (LR)**
- SVM
- NB
- **Multilayer perceptron (MLP)**
- **Decision tree (DT)**
- **Random forest (RF)**
- **Gradient boosted trees (GBT)**

For a classification problem, actual (that is, true) labels (that is, class) and the predicted label (that is, class) exist for the samples that are used to train or test a classifier; this can be assigned to one of the following categories:

- **True positive (TP)**: The true label is positive and the prediction made by the classifier is also positive
- **True negative (TN)**: The true label is negative and the prediction made by the classifier is also negative
- **False positive (FP)**: The true label is negative but the prediction made by the classifier is positive
- **False negative (FN)**: The true label is positive but the prediction made by the classifier is negative

These metrics (TP, FP, TN, and FN) are the building blocks of the evaluation metrics for most of the classifiers we listed previously. However, the pure accuracy that is often used for identifying how many predictions were correct is not generally a good metric, so other metrics such as precision, recall, F1 score, AUC, and **Matthew's correlation coefficient (MCC)** are used:

- *Accuracy* is the fraction of samples that the classifier correctly predicted (both positive and negative), divided by the total number of samples:

$$Accuracy = \frac{TP + TN}{TP + TN + FP + FN}$$

- *Precision* is the number of samples correctly predicted that belong to a positive class (true positives), divided by the total number of samples actually belonging to the positive class:

$$Precision = \frac{TP}{TP + FP}$$

- *Recall* is the number of samples that were correctly predicted to belong to a negative class, divided by the total number of elements actually belonging to the negative class:

$$Recall = \frac{TP}{TP + FN}$$

- F1 score is the harmonic mean of precision and recall. Since the F1 score is a balance between recall and precision, it can be considered as an alternative to accuracy:

$$F_1 = \frac{2 * Precision * Recall}{Precision + Recall}$$

Receiver Operating Characteristics (ROC) is a curve that's drawn by plotting **FPR** (to x-axis) and **TPR** (to y-axis) for different threshold values. So, for different thresholds for your classifier, we calculate the **TPR** and **FPR**, draw the **ROC** curve, and calculate the area under the **ROC** curve (also known as **AUC**). This can be visualized as follows:

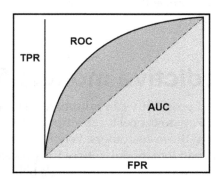

MCC is regarded as a balanced measure of a binary classifier, even for a dataset that has very imbalanced classes:

$$MCC = \frac{TP*TN - FP*FN}{\sqrt{(TP+FP)(TP+FN)(TN+FP)(TN+FN)}}$$

Let's discuss a more real-life example of a classification problem, which is churn analysis. Customer churn is the loss of clients or customers in any business, which is becoming a prime concern in different area of business, such as banks, internet service providers, insurance companies, and so on. Customer dissatisfaction and better offers from the competitor are the primary reasons behind this. In the telecommunications industry, when many subscribers switch to another service provider, the company not only loses those customers and revenue—this also creates a bad impression for other, regular customers, or people who were planning to start using their service.

Eventually, the full cost of customer churn includes both the lost revenue and the telemarketing costs involved with replacing those customers with new ones. However, these types of loss can cause a huge loss for a business. Remember the time when Nokia was the dominator in the cell phone market? All of a sudden, Apple announced iPhone 3G, which was a revolution in the smartphone era. Then, around 10% to 12% customers discontinued using Nokia and switched to iPhone. Although Nokia also tried to release a smartphone later on, they could not compete with Apple.

In short, churn prediction is essential for businesses as it helps you detect different kinds of customers who are likely to cancel a subscription, product, or service. In short, the idea is to predict whether an existing customer will unsubscribe from an existing service or not, that is, a binary classification problem.

Developing predictive models for churn

Accurate identification of churn possibility can minimize customer defection if you first identify which customers are likely to cancel a subscription to an existing service, and offering a special offer or plan to those customers. When it comes to employee churn prediction and developing a predictive model, where the process is heavily data-driven, machine learning can be used to understand a customer's behavior. This is done by analyzing the following:

- Demographic data, such as age, marital status, and job status
- Sentiment analysis based on their social media data

- Behavior analysis using their browsing clickstream logs
- Calling-circle data and support call center statistics

An automated churn analytics pipeline can be developed by following three steps:

1. First, identify typical tasks to analyze the churn, which will depend on company policy
2. Then, collect and analyze data and develop a predictive model
3. Finally, deploy the model in a production-ready environment

Eventually, telecom companies will be able to predict and enhance customer experience, prevent churn, and tailor marketing campaigns. In practice, such an analysis will be helpful to retain the customers who are most likely to leave. This means that we don't need to worry about the customers who are likely to stay.

Description of the dataset

We can use the Orange Telecom's Churn Dataset to develop a predictive model, which will predict which customers would like to cancel their subscription to existing services. The dataset is well studied, comprehensive, and used for developing small prototypes. It contains both churn-80 and churn-20 datasets, which can be downloaded from the following links:

- churn-80: `https://bml-data.s3.amazonaws.com/churn-bigml-80.csv`
- churn-20: `https://bml-data.s3.amazonaws.com/churn-bigml-20.csv`

Since both datasets came from the same distribution, which has the same structure, we will use the churn-80 dataset for the training and 10-fold cross-validation. Then, churn-20 will be used to evaluate the trained model. Both datasets have a similar structure, and therefore have the following schema:

- **State**: `String`
- **Account length**: `Integer`
- **Area code**: `Integer`
- **International plan**: `String`
- **Voicemail plan**: `String`
- **Number email messages**: `Integer`
- **Total day minutes**: `Double`
- **Total day calls**: `Integer`

- **Total day charge**: Double
- **Total evening minutes**: Double
- **Total evening calls**: Integer
- **Total evening charge**: Double
- **Total night minutes**: Double
- **Total night calls**: Integer
- **Total night charge**: Double
- **Total international minutes**: Double
- **Total international calls**: Integer
- **Total international charge**: Double
- **Customer service calls**: Integer

Exploratory analysis and feature engineering

First, we specify exactly the same schema (that is, a custom schema) before loading the data as a Spark DataFrame, as follows:

```
val schema = StructType(Array(
    StructField("state_code", StringType, true),
    StructField("account_length", IntegerType, true),
    StructField("area_code", StringType, true),
    StructField("international_plan", StringType, true),
    StructField("voice_mail_plan", StringType, true),
    StructField("num_voice_mail", DoubleType, true),
    StructField("total_day_mins", DoubleType, true),
    StructField("total_day_calls", DoubleType, true),
    StructField("total_day_charge", DoubleType, true),
    StructField("total_evening_mins", DoubleType, true),
    StructField("total_evening_calls", DoubleType, true),
    StructField("total_evening_charge", DoubleType, true),
    StructField("total_night_mins", DoubleType, true),
    StructField("total_night_calls", DoubleType, true),
    StructField("total_night_charge", DoubleType, true),
    StructField("total_international_mins", DoubleType, true),
    StructField("total_international_calls", DoubleType, true),
    StructField("total_international_charge", DoubleType, true),
    StructField("total_international_num_calls", DoubleType, true),
    StructField("churn", StringType, true)))
```

Then, we have to create a Scala case class with all the fields specified and align the preceding schema (variable names are self-explanatory):

```
case class CustomerAccount(state_code: String, account_length: Integer,
area_code: String,
                 international_plan: String, voice_mail_plan: String,
num_voice_mail: Double,
                 total_day_mins: Double, total_day_calls: Double,
total_day_charge: Double,
                 total_evening_mins: Double, total_evening_calls: Double,
total_evening_charge: Double,
                 total_night_mins: Double, total_night_calls: Double,
total_night_charge: Double,
                 total_international_mins: Double,
total_international_calls: Double,
                 total_international_charge: Double,
total_international_num_calls: Double, churn: String)
```

Let's create a Spark session and import the implicit._ package, which allows us to specify a DataFrame operation, as follows:

```
import spark.implicits._
```

Now, let's create the training set. We read the CSV file with Spark's recommended format, com.databricks.spark.csv. We do not need any explicit schema inference; hence, we are making the inferSchema false, but we are using our own schema, which we created previously. Then, we load the data file from our desired location, and finally specify our data source so that our DataFrame looks exactly the same as what we specified:

```
val trainSet: Dataset[CustomerAccount] = spark.read.
    option("inferSchema", "false")
    .format("com.databricks.spark.csv")
    .schema(schema)
    .load("data/churn-bigml-80.csv")
    .as[CustomerAccount]
trainSet.printSchema()
```

As we can see in the following screenshot, the schema of the Spark DataFrame has been correctly identified. However, some of the features are non-numeric but categorical. However, as expected by ML algorithms, all the features have to be numeric (that is, `integer` or `double` format):

```
root
 |-- state_code: string (nullable = true)
 |-- account_length: integer (nullable = true)
 |-- area_code: string (nullable = true)
 |-- international_plan: string (nullable = true)
 |-- voice_mail_plan: string (nullable = true)
 |-- num_voice_mail: double (nullable = true)
 |-- total_day_mins: double (nullable = true)
 |-- total_day_calls: double (nullable = true)
 |-- total_day_charge: double (nullable = true)
 |-- total_evening_mins: double (nullable = true)
 |-- total_evening_calls: double (nullable = true)
 |-- total_evening_charge: double (nullable = true)
 |-- total_night_mins: double (nullable = true)
 |-- total_night_calls: double (nullable = true)
 |-- total_night_charge: double (nullable = true)
 |-- total_international_mins: double (nullable = true)
 |-- total_international_calls: double (nullable = true)
 |-- total_international_charge: double (nullable = true)
 |-- total_international_num_calls: double (nullable = true)|
 |-- churn: string (nullable = true)
```

Excellent! It looks exactly the same as the data structure. Now, let's look at some sample data by using the `show()` method, as follows:

```
trainSet.show()
```

The output of the preceding line of code shows the first 20 samples of the DataFrame:

state	len	acode	intlplan	vplan	numvmail	tdmins	tdcalls	tdcharge	temins	tecalls	techarge	tnmins	tncalls	tncharge	timins	ticalls	ticharge	numcs	churn
KS	128	415	No	Yes	25.0	265.1	110.0	45.07	197.4	99.0	16.78	244.7	91.0	11.01	10.0	3.0	2.7	1.0	False
OH	107	415	No	Yes	26.0	161.6	123.0	27.47	195.5	103.0	16.62	254.4	103.0	11.45	13.7	3.0	3.7	1.0	False
NJ	137	415	No	No	0.0	243.4	114.0	41.38	121.2	110.0	10.3	162.6	104.0	7.32	12.2	5.0	3.29	0.0	False
OH	84	408	Yes	No	0.0	299.4	71.0	50.9	61.9	88.0	5.26	196.9	89.0	8.86	6.6	7.0	1.78	2.0	False
OK	75	415	Yes	No	0.0	166.7	113.0	28.34	148.3	122.0	12.61	186.9	121.0	8.41	10.1	3.0	2.73	3.0	False
AL	118	510	Yes	No	0.0	223.4	98.0	37.98	220.6	101.0	18.75	203.9	118.0	9.18	6.3	6.0	1.7	0.0	False
MA	121	510	No	Yes	24.0	218.2	88.0	37.09	348.5	108.0	29.62	212.6	118.0	9.57	7.5	7.0	2.03	3.0	False
MO	147	415	Yes	No	0.0	157.0	79.0	26.69	103.1	94.0	8.76	211.8	96.0	9.53	7.1	6.0	1.92	0.0	False
WV	141	415	Yes	Yes	37.0	258.6	84.0	43.96	222.0	111.0	18.87	326.4	97.0	14.69	11.2	5.0	3.02	0.0	False
RI	74	415	No	No	0.0	187.7	127.0	31.91	163.4	148.0	13.89	196.0	94.0	8.82	9.1	5.0	2.46	0.0	False
IA	168	408	No	No	0.0	128.8	96.0	21.9	104.9	71.0	8.92	141.1	128.0	6.35	11.2	2.0	3.02	1.0	False
MT	95	510	No	No	0.0	156.6	88.0	26.62	247.6	75.0	21.05	192.3	115.0	8.65	12.3	5.0	3.32	3.0	False
IA	62	415	No	No	0.0	120.7	70.0	20.52	307.2	76.0	26.11	203.0	99.0	9.14	13.1	6.0	3.54	4.0	False
ID	85	408	No	Yes	27.0	196.4	139.0	33.39	280.9	90.0	23.88	89.3	75.0	4.02	13.8	4.0	3.73	1.0	False
VT	93	510	No	No	0.0	190.7	114.0	32.42	218.2	111.0	18.55	129.6	121.0	5.83	8.1	3.0	2.19	3.0	False
VA	76	510	No	Yes	33.0	189.7	66.0	32.25	212.8	65.0	18.09	165.7	108.0	7.46	10.0	5.0	2.7	1.0	False
TX	73	415	No	No	0.0	224.4	90.0	38.15	159.5	88.0	13.56	192.8	74.0	8.68	13.0	2.0	3.51	1.0	False
FL	147	415	No	No	0.0	155.1	117.0	26.37	239.7	93.0	20.37	208.8	133.0	9.4	10.6	4.0	2.86	0.0	False
CO	77	408	No	No	0.0	62.4	89.0	10.61	169.9	121.0	14.44	209.6	64.0	9.43	5.7	6.0	1.54	5.0	True
AZ	130	415	No	No	0.0	183.0	112.0	31.11	72.9	99.0	6.2	181.8	78.0	8.18	9.5	19.0	2.57	0.0	False

only showing top 20 rows

In the preceding screenshot, column names are made shorter for visibility. We can also see related statistics of the training set by using the `describe()` method:

```
val statsDF = trainSet.describe()
statsDF.show()
```

The following summary statistics not only give us some idea on the distribution with mean and standard deviation of the data, but also some descriptive statistics, such as number samples (that is, count), minimum value, and maximum value for each feature in the DataFrame:

```
+-------+----------+------------------+------------------+-----------------+---------------+
|summary|state_code|    account_length|         area_code|international_plan|voice_mail_plan|
+-------+----------+------------------+------------------+-----------------+---------------+
|  count|      2666|              2666|              2666|             2666|           2666|
|   mean|      null|100.62040510127532|437.43885971492875|             null|           null|
| stddev|      null| 39.56397365334985|42.521018019427174|             null|           null|
|    min|        AK|                 1|               408|               No|             No|
|    max|        WY|               243|               510|              Yes|            Yes|
+-------+----------+------------------+------------------+-----------------+---------------+
```

If this dataset can fit into RAM, we can cache it for quick and repeated access by using the `cache()` method from Spark:

```
trainSet.cache()
```

Let's look at some useful properties such as variable correlation with `churn`. For example, let's see how the `churn` is related with total number of international calls:

```
trainSet.groupBy("churn").sum("total_international_num_calls").show()
```

As we can see from the following output, customers who make more international calls are less likely (that is, `False`) to change operator:

```
+-----+----------------------------------+
|churn|sum(total_international_num_calls)|
+-----+----------------------------------+
|False|                           3310.0|
|True |                            856.0|
+-----+----------------------------------+
```

Let's see how the `churn` is related to total international call charges:

```
trainSet.groupBy("churn").sum("total_international_charge").show()
```

As we can see from the following output, customers who make more international calls (as shown earlier) are charged more, but are still less likely (that is, `False`) to change operator:

```
+-----+-------------------------------+
|churn|sum(total_international_charge)|
+-----+-------------------------------+
|False|                 6236.499999999996|
| True|                           1133.63|
+-----+-------------------------------+
```

Now that we also need to have the test set prepared to evaluate the model, let's prepare the same set, similar to the train set, as follows:

```scala
val testSet: Dataset[CustomerAccount] = spark.read
    .option("inferSchema", "false")
    .format("com.databricks.spark.csv")
    .schema(schema)
    .load("data/churn-bigml-20.csv")
    .as[CustomerAccount]
```

Now, let's cache them for faster and quick access for further manipulation:

```scala
testSet.cache()
```

Let's look at some of the related properties of the training set to understand how suitable it is for our purposes. First, let's create a temp view for persistence for this session. Nevertheless, we can create a catalog as an interface that can be used to create, drop, alter, or query underlying databases, tables, functions, and so on:

```scala
trainSet.createOrReplaceTempView("UserAccount")
spark.catalog.cacheTable("UserAccount")
```

We can now group the data by the `churn` label and count the number of instances in each group, as follows:

```scala
trainSet.groupBy("churn").count.show()
```

The preceding line should show that only 388 customers are likely to switch to another operator. However, 2278 customers still have their current operator as their preferred one:

```
+-----+-----+
|churn|count|
+-----+-----+
|False| 2278|
| True| 388 |
+-----+-----+
```

So, we have roughly seven times more `False` churn samples than `True` churn samples. Since the target is to retain the customers who are most likely to leave, we will prepare our training set so that it ensures that the predictive ML model is sensitive to the `True` churn samples.

Also, since the training set is highly unbalanced, we should downsample the `False` churn class to a fraction of 388/2278, which gives us `0.1675`:

```
val fractions = Map("False" -> 0.1675, "True" -> 1.0)
```

This way, we are also mapping only `True` churn samples. Now, let's create a new DataFrame for the training set containing the samples from the downsample one only using the `sampleBy()` method:

```
val churnDF = trainSet.stat.sampleBy("churn", fractions, 12345L)
```

The third parameter is the seed that's used for reproducibility purposes. Let's take a look at this:

```
churnDF.groupBy("churn").count.show()
```

Now, we can see that the classes are almost balanced:

```
+-----+-----+
|churn|count|
+-----+-----+
|False|  390|
| True|  388|
+-----+-----+
```

Now, let's see how the variables are related to each other. Let's see how the day, night, evening, and international voice calls contribute to the `churn` class:

```
spark.sql()("SELECT churn, SUM(total_day_charge) as TDC,
                          SUM(total_evening_charge) as TEC,
SUM(total_night_charge) as TNC,
                          SUM(total_international_charge) as TIC,
                          SUM(total_day_charge) +
SUM(total_evening_charge) +
                          SUM(total_night_charge) +
SUM(total_international_charge)
                as Total_charge FROM UserAccount GROUP BY churn
                ORDER BY Total_charge DESC").show()
```

This, however, doesn't give any clear correlation because customers who are likely to stay make more day, night, evening, and international voice calls than the other customers who want to leave:

```
+-----+------------------+------------------+--------+------------------+------------------+
|churn|               TDC|               TEC|     TNC|               TIC|      Total_charge|
+-----+------------------+------------------+--------+------------------+------------------+
|False| 67812.10999999997|38504.60000000002|20549.78|6236.499999999996|         133102.99|
| True|13533.960000000005|6905.569999999997| 3584.69|           1133.63|25157.850000000002|
+-----+------------------+------------------+--------+------------------+------------------+
```

Now, let's see how many minutes of voice calls on day, night, evening, and international voice calls have contributed to the preceding `Total_charge` for the `churn` class:

```
spark.sql()("SELECT churn, SUM(total_day_mins) +
                    SUM(total_evening_mins) +
                    SUM(total_night_mins) +
                    SUM(total_international_mins) as Total_minutes
            FROM UserAccount GROUP BY churn")
        .show()
```

From the preceding two tables, it is clear that the total day minutes and total day charge are a highly correlated feature in this training set, which is not beneficial for our ML model training. Therefore, it would be better to remove them altogether. Let's drop one column of each pair of correlated fields, along with the `state_code` and `area_code` columns too since those won't be used:

```
val trainDF = churnDF
    .drop("state_code")
    .drop("area_code")
    .drop("voice_mail_plan")
    .drop("total_day_charge")
    .drop("total_evening_charge")
```

Excellent! Finally, we have our training DataFrame, which can be used for better, predictive modeling. Let's take a look at some of the columns of the resulting DataFrame:

```
trainDF.select("account_length", "international_plan", "num_voice_mail",
                "total_day_calls","total_international_num_calls", "churn")
        .show(10)
```

However, we are not done yet—the current DataFrame cannot be fed to the model. This is known as an estimator. As we described earlier, our data needs to be converted into a Spark DataFrame format consisting of labels (in `Double`) and features (in `Vector`):

```
+--------------+-----------------+--------------+---------------+---------------------------+-----+
|account_length|international_plan|num_voice_mail|total_day_calls|total_international_num_calls|churn|
+--------------+-----------------+--------------+---------------+---------------------------+-----+
|            84|              Yes|           0.0|           71.0|                        2.0|False|
|            74|               No|           0.0|          127.0|                        0.0|False|
|            95|               No|           0.0|           88.0|                        3.0|False|
|            62|               No|           0.0|           70.0|                        4.0|False|
|           147|               No|           0.0|          117.0|                        0.0|False|
|            77|               No|           0.0|           89.0|                        5.0| True|
|           130|               No|           0.0|          112.0|                        0.0|False|
|           142|               No|           0.0|           95.0|                        2.0|False|
|            12|               No|           0.0|          118.0|                        1.0| True|
|            57|               No|          25.0|           94.0|                        0.0|False|
+--------------+-----------------+--------------+---------------+---------------------------+-----+
only showing top 10 rows
```

Now, we need to create a pipeline to pass the data through by chaining several transformers and estimators. The pipeline then works as a feature extractor. More specifically, we have prepared two StringIndexer transformers and a VectorAssembler.

The first StringIndexer converts the String categorical feature international_plan and labels into number indices. The second StringIndexer converts the categorical label (that is, churn) into numeric format. This way, indexing categorical features allows decision trees and random forest-like classifiers to treat categorical features appropriately, thus improving performance:

```
val ipindexer = new StringIndexer()
      .setInputCol("international_plan")
      .setOutputCol("iplanIndex")

val labelindexer = new StringIndexer()
      .setInputCol("churn")
      .setOutputCol("label")
```

Now, we need to extract the most important features that contribute to the classification. Since we have dropped some columns already, the resulting column set consists of the following fields:

- Label → churn: True or False
- Features → {account_length, iplanIndex, num_voice_mail, total_day_mins, total_day_calls, total_evening_mins, total_evening_calls, total_night_mins, total_night_calls, total_international_mins, total_international_calls, total_intern ational_num_calls}

Since we have already converted categorical labels into numeric ones using `StringIndexer`, the next task is to extract the features:

```scala
val featureCols = Array("account_length", "iplanIndex", "num_voice_mail",
                        "total_day_mins", "total_day_calls",
"total_evening_mins",
                        "total_evening_calls", "total_night_mins",
"total_night_calls",
                        "total_international_mins",
"total_international_calls",
                        "total_international_num_calls")
```

Now, let's transform the features into feature vectors using `VectorAssembler()`, which takes all the `featureCols` and combines/transforms them into a single column called features:

```scala
val assembler = new VectorAssembler()
    .setInputCols(featureCols)
    .setOutputCol("features")
```

Now that we have the real training set, which consists of label and feature vectors ready, the next task is to create an estimator—the third element of a pipeline. We will start with a very simple but powerful LR classifier.

LR for churn prediction

LR is an algorithm for classification, which predicts a binary response. It is similar to linear regression, which we described in Chapter 2, *Scala for Regression Analysis*, except that it does not predict continuous values—it predicts discrete classes. The loss function is the sigmoid function (or logistic function):

$$L(w; x, y) := log(1 + exp(-yw^T x))$$

Similar to linear regression, the intuition behind the cost function is to penalize models that have large errors between the real response and the predicted response:

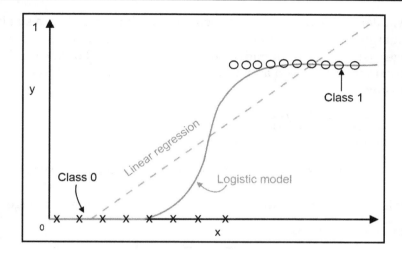

For a given new data point, **x**, the LR model makes predictions using the following equation:

$$f(z) = \frac{1}{1 + e^z}$$

In the preceding equation, the logistic function is applied to the regression to get the probabilities of it belonging in either class, where $z = w^T x$ and if $f(w^T x) > 0.5$, the outcome is positive; otherwise, it is negative. This means that the threshold for the classification line is assumed to be at *0.5*.

Now that we know how the LR algorithm works, let's start using Spark ML-based LR estimator development, which will predict whether a customer is likely to get churn or not. First, we need to define some hyperparameters to train a LR-based pipeline:

```
val numFolds = 10
val MaxIter: Seq[Int] = Seq(100)
val RegParam: Seq[Double] = Seq(1.0) // L2 regularization param set 1.0
with L1 reg. to reduce overfitting
val Tol: Seq[Double] = Seq(1e-8)// for convergence tolerance for iterative
algorithms
val ElasticNetParam: Seq[Double] = Seq(0.0001) //Combination of L1 & L2
```

RegParam is a scalar that helps us adjust the strength of the constraints: a small value implies a soft margin, while a large value implies a hard margin. The Tol parameter is used for the convergence tolerance for iterative algorithms, such as LR or linear SVM. Once we have the hyperparameters defined and initialized, our next task is to instantiate a LR estimator, as follows:

```scala
val lr = new LogisticRegression()
        .setLabelCol("label")
        .setFeaturesCol("features")
```

Now, let's build a pipeline estimator using the Pipeline() method to chain three transformers (the ipindexer, labelindexer, and assembler vectors) and the LR estimator (that is, lr) in a single pipeline—that is, each of them acts as a stage:

```scala
val pipeline = new Pipeline()
        .setStages(Array(PipelineConstruction.ipindexer,
                PipelineConstruction.labelindexer,
                    PipelineConstruction.assembler, lr))
```

A Spark ML pipeline can have the following components:

- **DataFrame**: To hold original data and intermediate transformed ones.

- **Transformer**: Used to transform one DataFrame into another by adding additional feature columns.
- **Estimator**: An estimator is an ML model, such as linear regression.
- **Pipeline**: Used to chain the preceding components, DataFrame, transformer, and estimator together.
- **Parameter**: An ML algorithm has many knobs to tweak. These are called hyperparameters, and the values learned by an ML algorithm to fit the data are called parameters.

In order to perform such a grid search over the hyperparameter space, we need to define it first. Here, the functional programming properties of Scala are quite handy because we just add function pointers and the respective parameters to be evaluated to the parameter grid. Here, a cross-validation evaluator will search through LR's max iteration, regularization param, tolerance, and elastic net for the best model:

```scala
val paramGrid = new ParamGridBuilder()
        .addGrid(lr.maxIter, MaxIter)
        .addGrid(lr.regParam, RegParam)
```

```
.addGrid(lr.tol, Tol)
.addGrid(lr.elasticNetParam, ElasticNetParam)
.build()
```

Note that the hyperparameters form an *n*-dimensional space, where *n* is the number of hyperparameters. Every point in this space is one particular hyperparameter configuration, which is a hyperparameter vector. Of course, we can't explore every point in this space, so what we basically do is a grid search over an (ideally evenly distributed) subset in that space. We then need to define a `BinaryClassificationEvaluator` evaluator, since this is a binary classification problem:

```
val evaluator = new BinaryClassificationEvaluator()
                .setLabelCol("label")
                .setRawPredictionCol("prediction")
```

We use a `CrossValidator` by using `ParamGridBuilder` to iterate through the max iteration, regression param, tolerance, and elastic net parameters of LR with 10-fold cross-validation:

```
val crossval = new CrossValidator()
    .setEstimator(pipeline)
    .setEvaluator(evaluator)
    .setEstimatorParamMaps(paramGrid)
    .setNumFolds(numFolds)
```

The preceding code is meant to perform cross-validation. The validator itself uses the `BinaryClassificationEvaluator` estimator to evaluate the training in the progressive grid space on each fold and make sure that no overfitting occurs.

Although there is so much stuff going on behind the scenes, the interface of our `CrossValidator` object stays slim and well known as `CrossValidator` also extends from the estimator and supports the `fit` method. This means that, after calling the `fit` method, the complete predefined pipeline, including all feature preprocessing and the LR classifier, is executed multiple times—each time with a different hyperparameter vector:

```
val cvModel = crossval.fit(Preprocessing.trainDF)
```

Now, it's time to evaluate the LR model using the test dataset. First, we need to transform the test set, similar to the training set we described previously:

```
val predDF= cvModel.transform(Preprocessing.testSet)
val result = predDF.select("label", "prediction", "probability")
val resutDF = result.withColumnRenamed("prediction", "Predicted_label")
resutDF.show(10)
```

The preceding code block shows the `Predicted_label` and the raw `probability` that were generated by the model. Additionally, it shows the actual labels. As we can see, for some instances, the model predicted correctly, but for other instances, it got confused:

```
+-----+---------------+--------------------+
|label|Predicted_label|         probability|
+-----+---------------+--------------------+
|  0.0|            0.0|[0.50128534704370...|
|  1.0|            0.0|[0.50128534704370...|
|  1.0|            0.0|[0.50128534704370...|
|  0.0|            0.0|[0.50128534704370...|
|  0.0|            0.0|[0.50128534704370...|
|  0.0|            0.0|[0.50128534704370...|
|  0.0|            0.0|[0.50128534704370...|
|  1.0|            0.0|[0.50128534704370...|
|  0.0|            0.0|[0.50128534704370...|
|  0.0|            0.0|[0.50128534704370...|
+-----+---------------+--------------------+
only showing top 10 rows
```

The prediction probabilities can also be very useful in ranking customers according to their likeliness of imperfection. This way, a limited number of resources can be utilized in telecommunication business that can be focused on the most valuable customers. However, by looking at the preceding prediction DataFrame, it is difficult to guess the classification accuracy. However, in the second step, the evaluator evaluates itself using `BinaryClassificationEvaluator`, as follows:

```
val accuracy = evaluator.evaluate(predDF)
println("Classification accuracy: " + accuracy)
```

This should show around 77% classification accuracy from our binary classification model:

```
Classification accuracy: 0.7679333824070667
```

We compute another performance metric called area under the precision-recall curve and the area under the ROC curve. For this, we can construct an RDD containing the raw scores on the test set:

```
val predictionAndLabels = predDF
      .select("prediction", "label")
      .rdd.map(x => (x(0).asInstanceOf[Double], x(1)
      .asInstanceOf[Double]))
```

Now, the preceding RDD can be used to compute the aforementioned performance metrics:

```
val metrics = new BinaryClassificationMetrics(predictionAndLabels)
println("Area under the precision-recall curve: " + metrics.areaUnderPR)
println("Area under the receiver operating characteristic (ROC) curve : " +
metrics.areaUnderROC)
```

In this case, the evaluation returns 77% accuracy, but only 58% precision:

```
Area under the precision-recall curve: 0.5770932703444629
Area under the receiver operating characteristic (ROC) curve:
0.7679333824070667
```

In the following code, we are calculating some more metrics. False and true positive and negative predictions are also useful to evaluate the model's performance. Then, we print the results to see the metrics, as follows:

```
val tVSpDF = predDF.select("label", "prediction") // True vs predicted
labels
val TC = predDF.count() //Total count

val tp = tVSpDF.filter($"prediction" === 0.0)
              .filter($"label" === $"prediction")
              .count() / TC.toDouble

val tn = tVSpDF.filter($"prediction" === 1.0)
              .filter($"label" === $"prediction")
              .count() / TC.toDouble

val fp = tVSpDF.filter($"prediction" === 1.0)
              .filter(not($"label" === $"prediction"))
              .count() / TC.toDouble
val fn = tVSpDF.filter($"prediction" === 0.0)
              .filter(not($"label" === $"prediction"))
              .count() / TC.toDouble
println("True positive rate: " + tp *100 + "%")
println("False positive rate: " + fp * 100 + "%")
println("True negative rate: " + tn * 100 + "%")
println("False negative rate: " + fn * 100 + "%")
```

The preceding code segment shows the true positive, false positive, true negative, and false negative rates, which we will use to compute the MCC score later on:

```
True positive rate: 66.71664167916042%
False positive rate: 19.04047976011994%
True negative rate: 10.944527736131935%
False negative rate: 3.2983508245877062%
```

Finally, we also compute the MCC score, as follows:

```
val MCC = (tp * tn - fp * fn) / math.sqrt((tp + fp) * (tp + fn) * (fp + tn)
* (tn + fn))
println("Matthews correlation coefficient: " + MCC)
```

The preceding line gives a Matthews correlation coefficient of 0.41676531680973805. This is a positive value, which gives us some sign of a robust model. However, we have not received good accuracy yet, so let's move on and try other classifiers, such as NB. This time, we will use the liner NB implementation from the Apache Spark ML package.

NB for churn prediction

The NB classifier is based on Bayes' theorem, with the following assumptions:

- Independence between every pair of features
- Feature values are non-negative, such as counts

For example, if cancer is related to age, this can be used to assess the probability that a patient might have cancer. Bayes' theorem is stated mathematically as follows:

$$\underset{\text{Posterior}}{P(A|B)} = \frac{\overset{\text{Likelihood Prior}}{P(B|A)\,P(A)}}{\underset{\text{Marginal likelihood}}{P(B)}}$$

In the preceding equation, A and B are events with $P(B) \neq 0$. The other terms can be described as follows:

- $P(A \mid B)$ is called the posterior or the conditional probability of observing event A, given that B is true
- $P(B \mid A)$ is the likelihood of event B given that A is true
- $P(A)$ is the prior and $P(B)$ is the prior probability, also called marginal likelihood or marginal probability

Gaussian NB is a generalized version of NB that's used for classification, which is based on the binomial distribution of data. For example, our churn prediction problem can be formulated as follows:

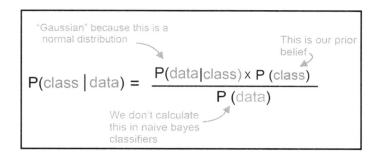

The preceding list can be adopted to solve our problem as follows:

- *P(class|data)* is the posterior probability of the *class* to be predicted by modelling with an independent variable (*data*)
- *P(data|class)* is the likelihood or the probability of the predictor, given *class*
- *P(class)* is the prior probability of *class* and *P(data)* of the predictor or marginal likelihood

The well-known Harvard study on happiness shows that only 10% of happy people are rich. Although you might think that this statistic is very compelling, you might be somewhat interested in knowing the percentage of rich people who are also really happy. Bayes' theorem helps you out with calculating this reserving statistic using two additional clues:

- The percentage of people overall who are happy—that is, *P(A)*
- The percentage of people overall who are rich—that is, *P(B)*

The key idea behind Bayes' theorem is reversing the statistic by considering the overall rates. Suppose the following pieces of information are available prior:

- 40% of people are happy => *P(A)*
- 5% of people are rich => *P(B)*

Now, let's consider that the Harvard study is correct—that is, *P(B|A)* = 10%. Now that we know the fraction of rich people who are happy, *P(A|B)* can be calculated as follows:

$$P(A|B) = \{P(A)* P(B|A)\} / P(B) = (40\%*10\%)/5\% = 80\%$$

Consequently, a majority of people are also happy! Nice. To make this clearer, let's assume that the population of the whole world is 5,000, for simplicity. According to our calculation, two facts exist:

- **Fact 1**: This tells us there are 500 people are happy, and the Harvard study tells us that 50 of these happy people are also rich
- **Fact 2**: There are 60 rich people altogether, so the fraction of them who are happy is 50/60 ~ 83%

This proves Bayes theorem and its effectiveness. To use NB, we need to instantiate an NB estimator, as follows:

```scala
val nb = new NaiveBayes()
        .setLabelCol("label")
        .setFeaturesCol("features")
```

Now that we have transformers and an estimator ready, the next task is to chain in a single pipeline—that is, each of them acts as a stage:

```scala
val pipeline = new Pipeline()
        .setStages(Array(PipelineConstruction.ipindexer,
            PipelineConstruction.labelindexer,
            PipelineConstruction.assembler,nb))
```

Let's define the `paramGrid` to perform such a grid search over the hyperparameter space. Then the cross-validator will search for the best model through the NB's `smoothing` parameter. Unlike LR or SVM, there is no hyperparameter in the NB algorithm:

```scala
val paramGrid = new ParamGridBuilder()
    .addGrid(nb.smoothing, Array(1.0, 0.1, 1e-2, 1e-4))// default value
is 1.0
    .build()
```

 Additive smoothing, or Laplace smoothing, is a technique that's used to smooth categorical data.

Let's define a `BinaryClassificationEvaluator` evaluator to evaluate the model:

```scala
val evaluator = new BinaryClassificationEvaluator()
                .setLabelCol("label")
                .setRawPredictionCol("prediction")
```

We use a `CrossValidator` to perform 10-fold cross-validation for best model selection:

```
val crossval = new CrossValidator()
    .setEstimator(pipeline)
    .setEvaluator(evaluator)
    .setEstimatorParamMaps(paramGrid)
    .setNumFolds(numFolds)
```

Let's call the `fit()` method so that the complete predefined `pipeline`, including all feature preprocessing and the LR classifier, is executed multiple times—each time with a different hyperparameter vector:

```
val cvModel = crossval.fit(Preprocessing.trainDF)
```

Now, it's time to evaluate the predictive power of the SVM model on the test dataset. First, we need to transform the test set with the model pipeline, which will map the features according to the same mechanism we described in the preceding feature engineering step:

```
val predDF = cvModel.transform(Preprocessing.testSet)
```

However, by looking at the preceding prediction DataFrame, it is difficult to guess the classification accuracy. However, in the second step, the evaluator evaluates itself using `BinaryClassificationEvaluator`, as follows:

```
val accuracy = evaluator.evaluate(predDF)
println("Classification accuracy: " + accuracy)
```

The preceding line of code should show 75% classification accuracy for our binary classification model:

```
Classification accuracy: 0.600772911299227
```

Like we did previously, we construct an RDD containing the raw scores on the test set:

```
val predictionAndLabels = predDF.select("prediction", "label")
    .rdd.map(x => (x(0).asInstanceOf[Double], x(1)
      .asInstanceOf[Double]))
```

Now, the preceding RDD can be used to compute the aforementioned performance metrics:

```
val metrics = new BinaryClassificationMetrics(predictionAndLabels)
println("Area under the precision-recall curve: " + metrics.areaUnderPR)
println("Area under the receiver operating characteristic (ROC) curve : " +
metrics.areaUnderROC)
```

In this case, the evaluation returns 75% accuracy but only 55% precision:

```
Area under the precision-recall curve: 0.44398397740763046
Area under the receiver operating characteristic (ROC) curve:
0.600772911299227
```

In the following code, again, we calculate some more metrics. False and true positive and negative predictions are also useful to evaluate the model's performance:

```
val tVSpDF = predDF.select("label", "prediction") // True vs predicted
labels
val TC = predDF.count() //Total count

val tp = tVSpDF.filter($"prediction" === 0.0)
              .filter($"label" === $"prediction")
              .count() / TC.toDouble

val tn = tVSpDF.filter($"prediction" === 1.0)
              .filter($"label" === $"prediction")
              .count() / TC.toDouble

val fp = tVSpDF.filter($"prediction" === 1.0)
              .filter(not($"label" === $"prediction"))
              .count() / TC.toDouble
val fn = tVSpDF.filter($"prediction" === 0.0)
              .filter(not($"label" === $"prediction"))
              .count() / TC.toDouble
println("True positive rate: " + tp *100 + "%")
println("False positive rate: " + fp * 100 + "%")
println("True negative rate: " + tn * 100 + "%")
println("False negative rate: " + fn * 100 + "%")
```

The preceding code segment shows the true positive, false positive, true negative, and false negative rates, which we will use to compute the MCC score later on:

```
True positive rate: 66.71664167916042%
False positive rate: 19.04047976011994%
True negative rate: 10.944527736131935%
False negative rate: 3.2983508245877062%
```

Finally, we also compute the MCC score, as follows:

```
val MCC = (tp * tn - fp * fn) / math.sqrt((tp + fp) * (tp + fn) * (fp + tn)
* (tn + fn))
println("Matthews correlation coefficient: " + MCC)
```

The preceding line gives a Matthews correlation coefficient of `0.14114315409796457` and this time, we experienced even worse performance in terms of accuracy and MCC scores. So, it's worth trying this with another classifier, such as SVM. We will use the linear SVM implementation from the Spark ML package.

SVM for churn prediction

SVM is also a population algorithm for classification. SVM is based on the concept of decision planes, which defines the decision boundaries we discussed at the beginning of this chapter. The following diagram shows how the SVM algorithm works:

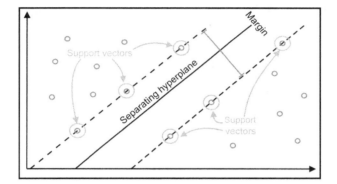

SVM uses kernel function, which finds the linear hyperplane that separates classes with the maximum margin. The following diagram shows how the data points (that is, support vectors) belonging to two different classes (red versus blue) are separated using the decision boundary based on the maximum margin:

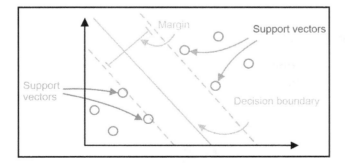

The preceding support vector classifier can be represented as a dot product mathematically, as follows:

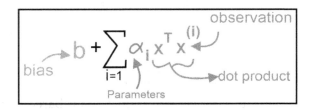

If the data to be separated is very high-dimensional, the kernel trick uses the kernel function to transform the data into a higher-dimensional feature space so that they can be linearly separable for classification. Mathematically, the kernel trick is to replace the dot product with the kernel, which will allow for non-linear decision boundaries and computational efficiency:

$$b + \sum \alpha_i k(x, x^{(i)}) \quad \text{Kernel}$$

Now that we already know about SVMs, let's start using the Spark-based implementation of SVM. First, we need to define some hyperparameters to train an LR-based pipeline:

```
val numFolds = 10
val MaxIter: Seq[Int] = Seq(100)
val RegParam: Seq[Double] = Seq(1.0) // L2 regularization param, set 0.10
with L1 regularization
val Tol: Seq[Double] = Seq(1e-8)
val ElasticNetParam: Seq[Double] = Seq(1.0) // Combination of L1 and L2
```

Once we have the hyperparameters defined and initialized, the next task is to instantiate an SVM estimator, as follows:

```
val svm = new LinearSVC()
```

Now that we have transformers and an estimator ready, the next task is to chain in a single pipeline—that is, each of them acts as a stage:

```
val pipeline = new Pipeline()
        .setStages(Array(PipelineConstruction.ipindexer,
            PipelineConstruction.labelindexer,
            PipelineConstruction.assembler, svm))
```

Let's define the `paramGrid` to perform such a grid search over the hyperparameter space. This searches through SVM's max iteration, regularization param, tolerance, and elastic net for the best model:

```
val paramGrid = new ParamGridBuilder()
      .addGrid(svm.maxIter, MaxIter)
      .addGrid(svm.regParam, RegParam)
      .addGrid(svm.tol, Tol)
      .addGrid(svm.elasticNetParam, ElasticNetParam)
      .build()
```

Let's define a `BinaryClassificationEvaluator` evaluator to evaluate the model:

```
val evaluator = new BinaryClassificationEvaluator()
            .setLabelCol("label")
            .setRawPredictionCol("prediction")
```

We use a `CrossValidator` to perform 10-fold cross-validation for best model selection:

```
val crossval = new CrossValidator()
      .setEstimator(pipeline)
      .setEvaluator(evaluator)
      .setEstimatorParamMaps(paramGrid)
      .setNumFolds(numFolds)
```

Now, let's call the fit method so that the complete predefined `pipeline`, including all feature preprocessing and the LR classifier, is executed multiple times—each time with a different hyperparameter vector:

```
val cvModel = crossval.fit(Preprocessing.trainDF)
```

Now, it's time to evaluate the predictive power of the SVM model on the test dataset:

```
val predDF= cvModel.transform(Preprocessing.testSet)
predDF.show(10)
```

The preceding code block shows the predicted label and the raw probability generated by the model. Additionally, it shows the actual labels.

As we can see, for some instances, the model predicted correctly, but for some other instances, it got confused:

```
+-----+---------------+--------------------+
|label|Predicted_label|         probability|
+-----+---------------+--------------------+
|  0.0|            1.0|[0.47967610185334...|
|  1.0|            1.0|[0.38531389528766...|
|  1.0|            1.0|[0.06850345623033...|
|  0.0|            0.0|[0.88942965019863...|
|  0.0|            0.0|[0.79162145165495...|
|  0.0|            0.0|[0.92841581163596...|
|  0.0|            1.0|[0.41083015977062...|
|  1.0|            1.0|[0.21531047960566...|
|  0.0|            1.0|[0.49987532729656...|
|  0.0|            1.0|[0.49313495545749...|
+-----+---------------+--------------------+
only showing top 10 rows
```

However, by looking at the preceding prediction DataFrame, it is difficult to guess the classification accuracy. However, in the second step, the evaluator evaluates itself using `BinaryClassificationEvaluator`, as follows:

```
val accuracy = evaluator.evaluate(predDF)
println("Classification accuracy: " + accuracy)
```

Therefore, we get about 75% classification accuracy from our binary classification model:

Classification accuracy: 0.7530180345969819

Now, we construct an RDD containing the raw scores on the test set, which will be used to compute performance metrics such as area under the precision-recall curve (AUC) and are under the received operating characteristic curve (ROC):

```
val predictionAndLabels = predDF
    .select("prediction", "label")
    .rdd.map(x => (x(0).asInstanceOf[Double], x(1)
    .asInstanceOf[Double]))
```

Now, the preceding RDD can be used to compute the aforementioned performance metrics:

```
val metrics = new BinaryClassificationMetrics(predictionAndLabels)
println("Area under the precision-recall curve: " + metrics.areaUnderPR)
println("Area under the receiver operating characteristic (ROC) curve : " +
metrics.areaUnderROC)
```

In this case, the evaluation returns 75% accuracy, but only 55% precision:

```
Area under the precision-recall curve: 0.5595712265324828
Area under the receiver operating characteristic (ROC) curve:
0.7530180345969819
```

We can also calculate some more metrics; for example, false and true positive and negative predictions are also useful to evaluate the model's performance:

```
val tVSpDF = predDF.select("label", "prediction") // True vs predicted
labels
val TC = predDF.count() //Total count

val tp = tVSpDF.filter($"prediction" === 0.0)
               .filter($"label" === $"prediction")
               .count() / TC.toDouble

val tn = tVSpDF.filter($"prediction" === 1.0)
               .filter($"label" === $"prediction")
               .count() / TC.toDouble

val fp = tVSpDF.filter($"prediction" === 1.0)
               .filter(not($"label" === $"prediction"))
               .count() / TC.toDouble
val fn = tVSpDF.filter($"prediction" === 0.0)
               .filter(not($"label" === $"prediction"))
               .count() / TC.toDouble
println("True positive rate: " + tp *100 + "%")
println("False positive rate: " + fp * 100 + "%")
println("True negative rate: " + tn * 100 + "%")
println("False negative rate: " + fn * 100 + "%")
```

The preceding code segment shows the true positive, false positive, true negative, and false negative rates, which we will use to compute the MCC score later on:

```
True positive rate: 66.71664167916042%
False positive rate: 19.04047976011994%
True negative rate: 10.944527736131935%
False negative rate: 3.2983508245877062%
```

Finally, we also compute the MCC score, as follows:

```
val MCC = (tp * tn - fp * fn) / math.sqrt((tp + fp) * (tp + fn) * (fp + tn)
* (tn + fn))
println("Matthews correlation coefficient: " + MCC)
```

This gave me a Matthews correlation coefficient of 0.3888239300421191. Although we have tried to use as many as three classification algorithms, we still haven't received good accuracy. Considering that SVM managed to give us an accuracy of 76%, this is still considered to be low. Moreover, there is no option for most suitable feature selection, which helps us train our model with the most appropriate features. To improve classification accuracy, we will need to use tree-based approaches, such as DT, RF, and GBT, which are expected to provide more powerful responses. We will do this in the next chapter.

Summary

In this chapter, we have learned about different classical classification algorithms, such as LR, SVM, and NB. Using these algorithms, we predicted whether a customer is likely to cancel their telecommunications subscription or not. We've also discussed what types of data are required to build a successful churn predictive model.

Tree-based and tree ensemble classifiers are really useful and robust, and are widely used for solving both classification and regression tasks. In the next chapter, we will look into developing such classifiers and regressors using tree-based and ensemble techniques such as DT, RF, and GBT, for both classification and regression.

4
Scala for Tree-Based Ensemble Techniques

In the previous chapter, we solved both classification and regression problems using linear models. We also used logistic regression, support vector machine, and Naive Bayes. However, in both cases, we haven't experienced good accuracy because our models showed low confidence.

On the other hand, tree-based and tree ensemble classifiers are really useful, robust, and widely used for both classification and regression tasks. This chapter will provide a quick glimpse at developing these classifiers and regressors using tree-based and ensemble techniques, such as **decision trees (DTs)**, **random forests (RF)**, and **gradient boosted trees (GBT)**, for both classification and regression. More specifically, we will revisit and solve both the regression (from Chapter 2, *Scala for Regression Analysis*) and classification (from Chapter 3, *Scala for Learning Classification*) problems we discussed previously.

The following topics will be covered in this chapter:

- Decision trees and tree ensembles
- Decision trees for supervised learning
- Gradient boosted trees for supervised learning
- Random forest for supervised learning
- What's next?

Technical requirements

Make sure Scala 2.11.x and Java 1.8.x are installed and configured on your machine.

The code files of this chapters can be found on GitHub:

`https://github.com/PacktPublishing/Machine-Learning-with-Scala-Quick-Start-Guide/tree/master/Chapter04`

Check out the following playlist to see the Code in Action video for this chapter:
`http://bit.ly/2WhQf2i`

Decision trees and tree ensembles

DTs normally fall under supervised learning techniques, which are used to identify and solve problems related to classification and regression. As the name indicates, DTs have various branches—where each branch indicates a possible decision, appearance, or reaction in terms of statistical probability. In terms of features, DTs are split into two main types: the training set and the test set, which helps produce a good update on the predicted labels or classes.

Both binary and multiclass classification problems can be handled by DT algorithms, which is one of the reasons it is used across problems. For instance, for the admission example we introduced in `Chapter 3`, *Scala for Learning Classification*, DTs learn from the admission data to approximate a sine curve with a set of `if...else` decision rules, as shown in the following diagram:

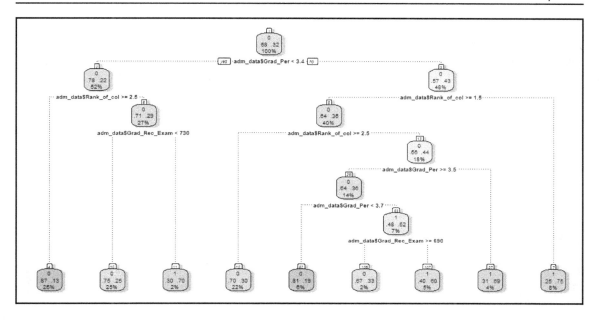

Generating decision rules using DTs based on university admission data

In general, the bigger the tree, the more complex the decision rules and the more fitted the model is. Another exciting power of DTs is that they can be used to solve both classification and regression problems. Now let's see some pros and cons of DTs. The two widely-used tree-based ensemble techniques are RF and GBT. The main difference between these two techniques is the way and order in which trees are trained:

- RFs train each tree independently but based on a random sample of the data. These random samples help to make the model more robust than a single DT, and hence it is less likely to have an overload on the training data.
- GBTs train one tree at a time. The errors created by the trees trained previously will be rectified by every new tree that is trained. As more trees are added, the model becomes more expressive.

RFs take a subset of observations and a subset of variables to build, which is an ensemble of DTs. These trees are actually trained on different parts of the same training set, but individual trees grow very deep tends to learn from highly unpredictable patterns.

 Sometimes very deep trees are responsible for overfitting problems in DT models. In addition, these biases can make the classifier a low performer even if the quality of the features represented is good with respect to the dataset.

When DTs are built, RFs integrate them together to get a more accurate and stable prediction. RFs helps to average multiple DTs together, with the goal of reducing the variance to ensure consistency by computing proximities between pairs of cases. This is a direct consequence on RF too. By maximum voting from a panel of independent jurors, we get the final prediction that is better than the best jury. The following figure shows how the decisions from two forests are ensembled together to get the final prediction:

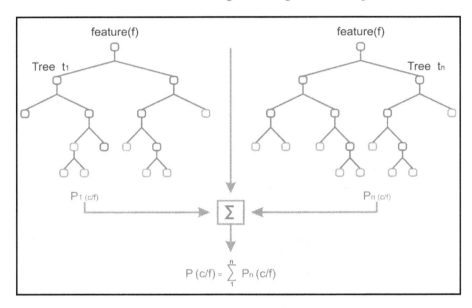

Tree-based ensemble and its assembling technique

In the end, both RF and GBT produce a weighted collection of DT, which is followed by predicting the combining results from the individual trees of each ensemble model. When using these approaches (as a classifier or regressor), the parameter settings are as follows:

- If the number of trees is 1, no bootstrapping is applied. If the number of trees is greater than 1, bootstrapping is applied, with `auto`, `all`, `sqrt`, `log2`, and one-third being the supported values.
- The supported numerical values are [0.0-1.0] and [1-n]. If `numTrees` is 1, `featureSubsetStrategy` is set to be `all`. If `numTrees` > 1 (for RF), `featureSubsetStrategy` is set to be `sqrt` for classification. If `featureSubsetStrategy` is chosen as `auto`, the algorithm infers the best feature subset strategy automatically.

- The impurity criterion is used only for the information-gain calculation, with gini and variance as the supported values for classification and regression, respectively.
- maxDepth is the maximum depth of the tree (for example, depth 0 means 1 leaf node, depth 1 means 1 internal node and 2 leaf nodes, and so on).
- maxBins signifies the maximum number of bins used to split the features, where the suggested value is 100 to get better results.

Now that we've already dealt with both regression analysis and classification problems, let's see how to use DT, RF, and GBT more comfortably to solve these problems. Let's get started with DT.

Decision trees for supervised learning

In this section, we'll see how to use DTs to solve both regression and classification problems. In the previous two chapters, Chapter 2, *Scala for Regression Analysis*, and Chapter 3, *Scala for Learning Classification*, we solved customer churn and insurance-severity claim problems. Those were classification and regression problems, respectively. In both approaches, we used other classic models. However, we'll see how we can solve them with tree-based and ensemble techniques. We'll use the DT implementation from the Apache Spark ML package in Scala.

Decision trees for classification

First of all, we know the customer churn prediction problem in Chapter 3, *Scala for Learning Classification*, and we know the data as well. We also know the working principle of DTs. So we can directly move to the coding part using the Spark based implementation of DTs. First we create a DecisionTreeClassifier estimator by instantiating the DecisionTreeClassifier interface. Additionally, we need to specify the label and feature vector columns:

```
val dTree = new DecisionTreeClassifier()
        .setLabelCol("label")// Setting label column
        .setFeaturesCol("features") // Setting feature vector column
        .setSeed(1234567L)// for reproducibility
```

As discussed in the previous chapters, we have three transformers (`ipindexer`, `labelindexer`, and `assembler`) and an estimator (`dTree`). We can now chain them together in a single pipeline so that each of them will act as a stage:

```
val pipeline = new Pipeline()
        .setStages(Array(PipelineConstruction.ipindexer,
                    PipelineConstruction.labelindexer,
                        PipelineConstruction.assembler,dTree))
```

Since we would like to perform hyperparameter tuning and cross-validation, we will have to create a `paramGrid` variable, which will be used for grid search over the hyperparameter space during the K-fold cross-validation:

```
var paramGrid = new ParamGridBuilder()
    .addGrid(dTree.impurity, "gini" :: "entropy" :: Nil)
    .addGrid(dTree.maxBins, 2 :: 5 :: 10 :: 15 :: 20 :: 25 :: 30 :: Nil)
    .addGrid(dTree.maxDepth, 5 :: 10 :: 15 :: 20 :: 25 :: 30 :: 30 :: Nil)
    .build()
```

More specifically, this will search for the DT's `impurity`, `maxBins`, and `maxDepth` for the best model. The maximum number of bins is used for separating continuous features and for choosing how to split on features at each node. Combined, the algorithm searches through the DT's `maxDepth` and `maxBins` parameters for the best model.

In the preceding code segment, we're creating a progressive `paramGrid` variable, where we specify the combination as a list of string or integer values. That means we are creating the grid space with different hyperparameter combinations. This will help us to provide the best model, consisting of optimal hyperparameters. However, for that, we need to have a `BinaryClassificationEvaluator` evaluator to evaluate each model and pick the best one during the cross-validation:

```
val evaluator = new BinaryClassificationEvaluator()
            .setLabelCol("label")
            .setRawPredictionCol("prediction")
```

We use `CrossValidator` to perform 10-fold cross validation to select the best model:

```
println("Preparing for 10-fold cross-validation")
val numFolds = 10

val crossval = new CrossValidator()
    .setEstimator(pipeline)
    .setEvaluator(evaluator)
    .setEstimatorParamMaps(paramGrid)
    .setNumFolds(numFolds)
```

Let's now call the `fit` method so that the complete predefined pipeline, including all feature preprocessing and the DT classifier, is executed multiple times—each time with a different hyperparameter vector:

```
val cvModel = crossval.fit(Preprocessing.trainDF)
```

Now it's time to evaluate the predictive power of the DT model on the test dataset:

```
val predictions = cvModel.transform(Preprocessing.testSet)
prediction.show(10)
```

This will lead us to the following DataFrame showing the predicted labels against the actual labels. Additionally, it shows the raw probabilities:

```
+-----+---------------+--------------------+
|label|Predicted_label|         probability|
+-----+---------------+--------------------+
|  0.0|            0.0|          [1.0,0.0]|
|  1.0|            1.0|[0.09523809523809...|
|  1.0|            1.0|          [0.0,1.0]||
|  0.0|            0.0|[0.86545454545454...|
|  0.0|            0.0|[0.86545454545454...|
|  0.0|            0.0|[0.86545454545454...|
|  0.0|            0.0|[0.80263157894736...|
|  1.0|            1.0|[0.09523809523809...|
|  0.0|            0.0|[0.86545454545454...|
|  0.0|            0.0|[0.86545454545454...|
+-----+---------------+--------------------+
only showing top 10 rows
```

However, based on the preceding prediction DataFrame, it is really difficult to guess the classification's accuracy. But in the second step, the evaluation is done using `BinaryClassificationEvaluator` as follows:

```
val accuracy = evaluator.evaluate(predictions)
println("Classification accuracy: " + accuracy)
```

This will provide an output with an accuracy value:

Accuracy: 0.8441663599558337

So, we get about 84% classification accuracy from our binary classification model. Just like with SVM and LR, we will observe the area under the precision-recall curve and the area under the **receiver operating characteristic (ROC)** curve based on the following RDD, which contains the raw scores on the test set:

```
val predictionAndLabels = predictions
        .select("prediction", "label")
        .rdd.map(x => (x(0).asInstanceOf[Double], x(1)
        .asInstanceOf[Double]))
```

The preceding RDD can be used for computing the previously mentioned two performance metrics:

```
val metrics = new BinaryClassificationMetrics(predictionAndLabels)
println("Area under the precision-recall curve: " + metrics.areaUnderPR)
println("Area under the receiver operating characteristic (ROC) curve: " +
metrics.areaUnderROC)
```

In this case, the evaluation returns 84% accuracy but only 67% precision, which is much better than that of SVM and LR:

```
Area under the precision-recall curve: 0.6665988000794282
Area under the receiver operating characteristic (ROC) curve:
0.8441663599558337
```

Then, we calculate some more metrics, for example, false and true positive, and false and true negative, as these predictions are also useful to evaluate the model's performance:

```
val TC = predDF.count() //Total count

val tp = tVSpDF.filter($"prediction" === 0.0).filter($"label" ===
$"prediction")
                    .count() / TC.toDouble // True positive rate
val tn = tVSpDF.filter($"prediction" === 1.0).filter($"label" ===
$"prediction")
                    .count() / TC.toDouble // True negative rate
val fp = tVSpDF.filter($"prediction" === 1.0).filter(not($"label" ===
$"prediction"))
                    .count() / TC.toDouble // False positive rate
val fn = tVSpDF.filter($"prediction" === 0.0).filter(not($"label" ===
$"prediction"))
                    .count() / TC.toDouble // False negative rate
```

Additionally, we compute the Matthews correlation coefficient:

```
val MCC = (tp * tn - fp * fn) / math.sqrt((tp + fp) * (tp + fn) * (fp + tn)
* (tn + fn))
```

Let's observe how high the model confidence is:

```
println("True positive rate: " + tp *100 + "%")
println("False positive rate: " + fp * 100 + "%")
println("True negative rate: " + tn * 100 + "%")
println("False negative rate: " + fn * 100 + "%")
println("Matthews correlation coefficient: " + MCC)
```

Fantastic! We achieved only 70% accuracy, which is probably why we had a low number of trees, but for what factors?

```
True positive rate: 70.76461769115441%
False positive rate: 14.992503748125937%
True negative rate: 12.293853073463268%
False negative rate: 1.9490254872563717%
Matthews correlation coefficient: 0.5400720075807806
```

Now let's see at what level we achieved the best model after the cross-validation:

```
val bestModel = cvModel.bestModel
println("The Best Model and Parameters:\n--------------------")
println(bestModel.asInstanceOf[org.apache.spark.ml.PipelineModel].stages(3)
)
```

According to the following output, we achieved the best tree model at depth 5 with 53 nodes:

```
The Best Model and Parameters:
DecisionTreeClassificationModel of depth 5 with 53 nodes
```

Let's extract those moves taken (that is, decisions) during tree construction by showing the tree. This tree helps us to find the most valuable features in our dataset:

```
bestModel.asInstanceOf[org.apache.spark.ml.PipelineModel]
    .stages(3)
    .extractParamMap
val treeModel = bestModel.asInstanceOf[org.apache.spark.ml.PipelineModel]
    .stages(3)
    .asInstanceOf[DecisionTreeClassificationModel]
println("Learned classification tree model:\n" + treeModel.toDebugString)
```

In the following output, the `toDebugString()` method prints the tree's decision nodes and final the prediction outcomes at the end leaves:

```
Learned classification tree model:
If (feature 3 <= 245.2)
    If (feature 11 <= 3.0)
     If (feature 1 in {1.0})
      If (feature 10 <= 2.0)
       Predict: 1.0
      Else (feature 10 > 2.0)
       If (feature 9 <= 12.9)
        Predict: 0.0
       Else (feature 9 > 12.9)
        Predict: 1.0
     ...
  Else (feature 7 > 198.0)
      If (feature 2 <= 28.0)
       Predict: 1.0
      Else (feature 2 > 28.0)
       If (feature 0 <= 60.0)
        Predict: 0.0
       Else (feature 0 > 60.0)
        Predict: 1.0
```

We can also see that certain features (3 and 11 in our case) are used for decision making—that is, the two most important reasons customers are likely to churn. But what are those two features? Let's see them:

```
println("Feature 11:" +
Preprocessing.trainDF.filter(PipelineConstruction.featureCols(11)))
println("Feature 3:" +
Preprocessing.trainDF.filter(PipelineConstruction.featureCols(3)))
```

According to the following output, feature 3 and 11 were most important predictors:

```
Feature 11: [total_international_num_calls: double]
Feature 3:  [total_day_mins: double]
```

The customer service calls and total day minutes are selected by DTs, since they provide an automated mechanism for determining the most important features.

Decision trees for regression

In Chapter 3, *Scala for Learning Classification*, we learned how to predict the problem regarding slowness in traffic. We applied **linear regression** (**LR**) and generalized linear regression to solve this problem. Also, we knew the data very well.

As stated earlier, DT also can provide very powerful responses and performance in the case of a regression problem. Similar to DecisionTreeClassifier, a DecisionTreeRegressor estimator can be instantiated with the DecisionTreeRegressor() method. Additionally, we need to explicitly specify the label and feature columns:

```
// Estimator algorithm
val model = new
DecisionTreeRegressor().setFeaturesCol("features").setLabelCol("label")
```

We can set the max bins, number of trees, max depth, and impurity while instantiating the preceding estimator.

However, since we'll perform k-fold cross-validation, we can set those parameters while creating paramGrid:

```
// Search through decision tree's parameter for the best model
var paramGrid = new ParamGridBuilder()
        .addGrid(rfModel.impurity, "variance" :: Nil)// variance for
regression
        .addGrid(rfModel.maxBins, 25 :: 30 :: 35 :: Nil)
        .addGrid(rfModel.maxDepth, 5 :: 10 :: 15 :: Nil)
        .addGrid(rfModel.numTrees, 3 :: 5 :: 10 :: 15 :: Nil)
        .build()
```

For a better and more stable performance, let's prepare the k-fold cross-validation and grid search as a part of model tuning. As you can guess, I am going to perform 10-fold cross-validation. Feel free to adjust number of folds based on your settings and dataset:

```
println("Preparing K-fold Cross Validation and Grid Search: Model tuning")
val numFolds = 10  // 10-fold cross-validation
val cv = new CrossValidator()
        .setEstimator(rfModel)
        .setEvaluator(new RegressionEvaluator)
        .setEstimatorParamMaps(paramGrid)
        .setNumFolds(numFolds)
```

Fantastic! We have created the cross-validation estimator. Now it's time to train DT regression model with cross-validation:

```
println("Training model with decision tree algorithm")
val cvModel = cv.fit(trainingData)
```

Now that we have the fitted model, we can make predictions. So let's start evaluating the model on the train and validation set and calculate RMSE, MSE, MAE, R squared, and so on:

```
println("Evaluating the model on the test set and calculating the
regression metrics")
val trainPredictionsAndLabels = cvModel.transform(testData).select("label",
"prediction")
                                        .map { case Row(label: Double,
prediction: Double)
                                        => (label, prediction) }.rdd

val testRegressionMetrics = new
RegressionMetrics(trainPredictionsAndLabels)
```

Once we have the best-fitted and cross-validated model, we can expect a good prediction accuracy. Let's observe the result on the train and the validation set:

```
val results =
"\n=====================================================================\n"
+
      s"TrainingData count: ${trainingData.count}\n" +
      s"TestData count: ${testData.count}\n" +
"=====================================================================\n" +
      s"TestData MSE = ${testRegressionMetrics.meanSquaredError}\n" +
      s"TestData RMSE = ${testRegressionMetrics.rootMeanSquaredError}\n" +
      s"TestData R-squared = ${testRegressionMetrics.r2}\n" +
      s"TestData MAE = ${testRegressionMetrics.meanAbsoluteError}\n" +
      s"TestData explained variance =
${testRegressionMetrics.explainedVariance}\n" +
"=====================================================================\n"
println(results)
```

The following output shows the MSE, RMSE, R-squared, MAE and explained variance on the test set:

```
=====================================================================
TrainingData count: 80
TestData count: 55
=====================================================================
TestData MSE = 7.871519100933004
TestData RMSE = 2.8056227652578323
```

```
TestData R-squared = 0.5363607928629964
TestData MAE = 2.284866391184572
TestData explained variance = 20.213067468774792
====================================================================
```

Great! We have managed to compute the raw prediction on the train set and the test set, and we can see the improvements compared to LR regression model. Let's hunt for the model that will help us to achieve better accuracy:

```
val bestModel = cvModel.bestModel.asInstanceOf[DecisionTreeRegressionModel]
```

Additionally, we can see how the decisions were made by observing DTs in the forest:

```
println("Decision tree from best cross-validated model: " +
bestModel.toDebugString)
```

The following is the output:

```
Decision tree from best cross-validated model at depth 5 with 39 nodes
  If (feature 0 <= 19.0)
   If (feature 0 <= 3.0)
    If (feature 0 <= 1.0)
     If (feature 3 <= 0.0)
      If (feature 4 <= 0.0)
       Predict: 4.1
      Else (feature 4 > 0.0)
       Predict: 3.4000000000000004
     . . . .
       Predict: 15.30909090909091
      Else (feature 0 > 25.0)
       Predict: 12.800000000000002
     Else (feature 11 > 1.0)
      Predict: 22.100000000000023
   Else (feature 9 > 1.0)
    Predict: 23.399999999999977
```

With DTs, it is possible to measure the feature importance, so that in a later stage we can decide which features to use and which ones to drop from the DataFrame. Let's find the feature importance out of the best model we just created before for all the features that are arranged in an ascending order as follows:

```
val featureImportances = bestModel.featureImportances.toArray

val FI_to_List_sorted = featureImportances.toList.sorted.toArray
println("Feature importance generated by the best model: ")
for(x <- FI_to_List_sorted) println(x)
```

Following is the feature importance generated by the model:

```
Feature importance generated by the best model:
 0.0
 0.0
 0.0
 0.0
 0.0
 0.0
 0.0
 0.0
 7.109215735617604E-5
 2.1327647206851872E-4
 0.001134987328520092
 0.00418143999334111
 0.025448271970345014
 0.03446268498009088
 0.057588305610674816
 0.07952108027588178
 0.7973788612117217
```

The last result is important to understand the feature importance. As you can see, RF has ranked some features to be more important. For example, the last few features are the most important ones, while eight of them are less important. We can drop those unimportant columns and train the DT model again to observe whether there is any greater reduction of MAE and increase in R-squared on the test set.

Gradient boosted trees for supervised learning

In this section, we'll see how to use GBT to solve both regression and classification problems. In the previous two chapters, Chapter 2, *Scala for Regression Analysis*, and Chapter 3, *Scala for Learning Classification*, we solved the customer churn and insurance severity claim problems, which were classification and regression problem, respectively. In both approaches, we used other classic models. However, we'll see how we can solve them with tree-based and ensemble techniques. We'll use the GBT implementation from the Spark ML package in Scala.

Gradient boosted trees for classification

We know the customer churn prediction problem from Chapter 3, *Scala for Learning Classification*, and we know the data well. We already know the working principles of RF, so let's start using the Spark-based implementation of RF:

1. Instantiate a GBTClassifier estimator by invoking the GBTClassifier() interface:

```
val gbt = new GBTClassifier()
    .setLabelCol("label")
    .setFeaturesCol("features")
    .setSeed(1234567L)
```

2. We have three transformers and an estimator ready. Chain in a single pipeline, that is, each of them acts a stage:

```
// Chain indexers and tree in a Pipeline.
val pipeline = new Pipeline()
.setStages(Array(ScalaClassification.PipelineConstruction.ipindexer
,
        ScalaClassification.PipelineConstruction.labelindexer,
        ScalaClassification.PipelineConstruction.assembler,
        gbt))
```

3. Define the paramGrid variable to perform such a grid search over the hyperparameter space:

```
// Search through decision tree's maxDepth parameter for best model
val paramGrid = new ParamGridBuilder()
        .addGrid(gbt.maxDepth, 3 :: 5 :: 10 :: Nil) // :: 15 :: 20 ::
25 :: 30 :: Nil)
        .addGrid(gbt.impurity, "gini" :: "entropy" :: Nil)
        .addGrid(gbt.maxBins, 5 :: 10 :: 20 :: Nil) //10 :: 15 :: 25
:: 35 :: 45 :: Nil)
        .build()
```

4. Define a BinaryClassificationEvaluator evaluator to evaluate the model:

```
val evaluator = new BinaryClassificationEvaluator()
                .setLabelCol("label")
                .setRawPredictionCol("prediction")
```

5. We use a `CrossValidator` for performing 10-fold cross validation for best model selection:

```
// Set up 10-fold cross validation
val numFolds = 10
val crossval = new CrossValidator()
        .setEstimator(pipeline)
        .setEvaluator(evaluator)
        .setEstimatorParamMaps(paramGrid)
        .setNumFolds(numFolds)
```

6. Let's call now the `fit` method so that the complete predefined pipeline, including all feature preprocessing and DT classifier, is executed multiple times—each time with a different hyperparameter vector:

```
val cvModel = crossval.fit(Preprocessing.trainDF)
```

Now it's time to evaluate the predictive power of DT model on the test dataset:

1. Transform the test set with the model pipeline, which will update the features as per the same mechanism we described in the preceding feature engineering step:

```
val predictions = cvModel.transform(Preprocessing.testSet)
prediction.show(10)
```

This will lead us to the following DataFrame showing the predicted labels against the actual labels. Additionally, it shows the raw probabilities:

```
+-----+---------------+--------------------+
|label|Predicted_label|         probability|
+-----+---------------+--------------------+
|  0.0|            1.0|[0.47967610185334...|
|  1.0|            1.0|[0.38531389528766...|
|  1.0|            1.0|[0.06850345623033...|
|  0.0|            0.0|[0.88942965019863...|
|  0.0|            0.0|[0.79162145165495...|
|  0.0|            0.0|[0.92841581163596...|
|  0.0|            1.0|[0.41083015977062...|
|  1.0|            1.0|[0.21531047960566...|
|  0.0|            1.0|[0.49987532729656...|
|  0.0|            1.0|[0.49313495545749...|
+-----+---------------+--------------------+
only showing top 10 rows
```

However, after seeing the preceding prediction DataFrame, it is really difficult to guess the classification accuracy.

2. But in the second step, in the evaluation is done using `BinaryClassificationEvaluator` as follows:

```
val accuracy = evaluator.evaluate(predictions)
println("Classification accuracy: " + accuracy)
```

This will give us the classification accuracy:

Accuracy: 0.869460802355539

So we get about 87% classification accuracy from our binary classification model. Just like with SVM and LR, we will observe the area under the precision-recall curve and the area under the ROC curve based on the following RDD, which contains the raw scores on the test set:

```
val predictionAndLabels = predictions
        .select("prediction", "label")
        .rdd.map(x => (x(0).asInstanceOf[Double], x(1)
        .asInstanceOf[Double]))
```

The preceding RDD can be used for computing the previously mentioned performance metrics:

```
val metrics = new BinaryClassificationMetrics(predictionAndLabels)
println("Area under the precision-recall curve: " + metrics.areaUnderPR)
println("Area under the receiver operating characteristic (ROC) curve: " +
metrics.areaUnderROC)
```

This will share the value in terms of accuracy and prediction:

Area under the precision-recall curve: 0.7270259009251356
Area under the receiver operating characteristic (ROC) curve:
0.869460802355539

In this case, the evaluation returns 87% accuracy but only 73% precision, which is much better than SVM and LR. Then we calculate some more false and true metrics. Positive and negative predictions can also be useful to evaluate the model's performance:

```
val TC = predDF.count() //Total count

val tp = tVSpDF.filter($"prediction" === 0.0).filter($"label" ===
$"prediction")
                    .count() / TC.toDouble // True positive rate
val tn = tVSpDF.filter($"prediction" === 1.0).filter($"label" ===
```

```
$"prediction")
                       .count() / TC.toDouble // True negative rate
val fp = tVSpDF.filter($"prediction" === 1.0).filter(not($"label" ===
$"prediction"))
                       .count() / TC.toDouble // False positive rate
val fn = tVSpDF.filter($"prediction" === 0.0).filter(not($"label" ===
$"prediction"))
                       .count() / TC.toDouble // False negative rate
```

Additionally, we compute the Matthews correlation coefficient:

```
val MCC = (tp * tn - fp * fn) / math.sqrt((tp + fp) * (tp + fn) * (fp + tn)
* (tn + fn))
```

Let's observe how high the model confidence is:

```
println("True positive rate: " + tp *100 + "%")
println("False positive rate: " + fp * 100 + "%")
println("True negative rate: " + tn * 100 + "%")
println("False negative rate: " + fn * 100 + "%")
println("Matthews correlation coefficient: " + MCC)
```

Now let's take a look at the true positive, false positive, true negative, and false negative rates. Additionally, we see the MCC:

```
True positive rate: 0.7781109445277361
False positive rate: 0.07946026986506746
True negative rate: 0.1184407796101949
False negative rate: 0.0239880059970015
Matthews correlation coefficient: 0.6481780577821629
```

These rates looks promising as we experienced positive MCC that shows mostly positive correlation indicating a robust classifier. Now, similar to DTs, RFs can be debugged during the classification. For the tree to be printed and to select the most important features, run the last few lines of code in the DT. Note that we still confine the hyperparameter space with `numTrees`, `maxBins`, and `maxDepth` by limiting them to 7. Remember that bigger trees will most likely perform better. Therefore, feel free to play around with this code and add features, and also use a bigger hyperparameter space, for instance, bigger trees.

GBTs for regression

To reduce the size of a loss function, GBTs will train many DTs. For each instance, the algorithm will use the ensemble that is currently available to predict the label of each training.

Similar to decision trees, GBTs can do the following:

- Handle both categorical and numerical features
- Be used for both binary classification and regression (multiclass classification is not yet supported)
- Do not require feature scaling
- Capture non-linearity and feature interactions from very high-dimensional datasets

Suppose we have N data instances (being x_i = features of instance i) and y is the label (being y_i = label of instance i), then $f(x_i)$ is GBT model's predicted label for instance i, which tries to minimize any of the following losses:

$$L_{log} = 2\sum_{i=1}^{N} log(1 + exp(-2y_i f(xi)))\ldots\ldots(1)$$

$$L_{se} = \sum_{i=1}^{N} (y_i - f(x_i))^2\ldots\ldots(2)$$

$$L_{ae} = \sum_{i=1}^{N} |y_i - f(x_i)|\ldots\ldots(3)$$

The first equation is called the *log* loss, which is twice the binomial negative *log* likelihood. The second one called squared error is commonly referred to as *L2* loss and the default loss for GBT-based regression task. Finally, the third, called absolute error, is commonly referred to as *L1* loss and is recommended if the data points have many outliers and robust than squared error.

Now that we know the minimum working principle of the GBT regression algorithm, we can get started. Let's instantiate a GBTRegressor estimator by invoking the GBTRegressor() interface:

```
val gbtModel = new
GBTRegressor().setFeaturesCol("features").setLabelCol("label")
```

We can set the max bins, number of trees, max depth, and impurity when instantiating the preceding estimator. However, since we'll perform k-fold cross-validation, we can set those parameters while creating the paramGrid variable too:

```
// Search through GBT's parameter for the best model
var paramGrid = new ParamGridBuilder()
        .addGrid(gbtModel.impurity, "variance" :: Nil)// variance for
regression
        .addGrid(gbtModel.maxBins, 25 :: 30 :: 35 :: Nil)
```

```
.addGrid(gbtModel.maxDepth, 5 :: 10 :: 15 :: Nil)
.addGrid(gbtModel.numTrees, 3 :: 5 :: 10 :: 15 :: Nil)
.build()
```

 Validation while training: Gradient boosting can overfit, especially when you train your model with more trees. In order to prevent this issue, it is useful to validate (for example, using cross-validation) while carrying out the training.

For a better and more stable performance, let's prepare the k-fold cross-validation and grid search as part of the model tuning. As you can guess, I am going to perform 10-fold cross-validation. Feel free to adjust the number of folds based on your settings and dataset:

```
println("Preparing K-fold Cross Validation and Grid Search: Model tuning")
val numFolds = 10  // 10-fold cross-validation
val cv = new CrossValidator()
        .setEstimator(gbtModel)
        .setEvaluator(new RegressionEvaluator)
        .setEstimatorParamMaps(paramGrid)
        .setNumFolds(numFolds)
```

Fantastic! We have created the cross-validation estimator. Now it's time to train the RandomForestRegression model with cross-validation:

```
println("Training model with RandomForestRegressor algorithm")
val cvModel = cv.fit(trainingData)
```

Now that we have the fitted model, we can make predictions. Let's start evaluating the model on the train and validation sets and calculate RMSE, MSE, MAE, and R squared error:

```
println("Evaluating the model on the test set and calculating the
regression metrics")
val trainPredictionsAndLabels = cvModel.transform(testData).select("label",
"prediction")
                                    .map { case Row(label: Double,
prediction: Double)
                                    => (label, prediction) }.rdd

val testRegressionMetrics = new
RegressionMetrics(trainPredictionsAndLabels)
```

Once we have the best-fitted and cross-validated model, we can expect a high prediction accuracy. Now let's observe the result on the train and the validation sets:

```
val results =
"\n=====================================================================\n"
+
        s"TrainingData count: ${trainingData.count}\n" +
        s"TestData count: ${testData.count}\n" +
"=====================================================================\n" +
        s"TestData MSE = ${testRegressionMetrics.meanSquaredError}\n" +
        s"TestData RMSE = ${testRegressionMetrics.rootMeanSquaredError}\n" +
        s"TestData R-squared = ${testRegressionMetrics.r2}\n" +
        s"TestData MAE = ${testRegressionMetrics.meanAbsoluteError}\n" +
        s"TestData explained variance =
${testRegressionMetrics.explainedVariance}\n" +
"=====================================================================\n"
println(results)
```

The following output shows the MSE, RMSE, R-squared, MAE and explained variance on the test set:

```
=====================================================================
TrainingData count: 80
TestData count: 55
=====================================================================
TestData MSE = 5.99847335425882
TestData RMSE = 2.4491780977011084
TestData R-squared = 0.4223425609926217
TestData MAE = 2.0564380367107646
TestData explained variance = 20.340666319995183
=====================================================================
```

Great! We have managed to compute the raw prediction on the train and the test set, and we can see the improvements compared to the LR, DT, and GBT regression models. Let's hunt for the model that helps us to achieve better accuracy:

```
val bestModel = cvModel.bestModel.asInstanceOf[GBTRegressionModel]
```

Additionally, we can see how the decisions were made by observing the DTs in the forest:

```
println("Decision tree from best cross-validated model: " +
bestModel.toDebugString)
```

In the following output, the toDebugString() method prints the tree's decision nodes and final prediction outcomes at the end leaves:

```
Decision tree from best cross-validated model with 10 trees
  Tree 0 (weight 1.0):
    If (feature 0 <= 16.0)
     If (feature 2 <= 1.0)
      If (feature 15 <= 0.0)
       If (feature 13 <= 0.0)
        If (feature 16 <= 0.0)
         If (feature 0 <= 3.0)
          If (feature 3 <= 0.0)
           Predict: 6.128571428571427
          Else (feature 3 > 0.0)
           Predict: 3.3999999999999986
  . . . .
  Tree 9 (weight 1.0):
    If (feature 0 <= 22.0)
     If (feature 2 <= 1.0)
      If (feature 1 <= 1.0)
       If (feature 0 <= 1.0)
        Predict: 3.4
  . . .
```

With random forest, it is possible to measure the feature importance so that in a later stage, we can decide which features to use and which ones to drop from the DataFrame. Let's find the feature importance out of the best model we just created before for all the features that are arranged in an ascending order as follows:

```
val featureImportances = bestModel.featureImportances.toArray

val FI_to_List_sorted = featureImportances.toList.sorted.toArray
println("Feature importance generated by the best model: ")
for(x <- FI_to_List_sorted) println(x)
```

Following is the feature importance generated by the model:

```
Feature importance generated by the best model:
 0.0
 0.0
 5.767724652714395E-4
 0.001616872851121874
 0.006381209526062637
 0.008867810069950395
 0.009420668763121653
 0.01802097742361489
 0.0267557383338777407
 0.02761531441902482
```

```
0.031208534172407782
0.033620224027091
0.03801721834820778
0.05263475066123412
0.05562565266841311
0.13221209076999635
0.5574261654957049
```

The last result is important to understand the feature importance. As you can see, the RF has ranked some features that looks to be more important. For example, the last two features are the most important and the first two are less important. We can drop some unimportant columns and train the RF model to observe whether there is any reduction in the R-squared and MAE values on the test set.

Random forest for supervised learning

In this section, we'll see how to use RF to solve both regression and classification problems. We'll use DT implementation from the Spark ML package in Scala. Although both GBT and RF are ensembles of trees, the training processes are different. For instance, RF uses the bagging technique to perform the example, while GBT uses boosting. Nevertheless, there are several practical trade-offs between both the ensembles that can pose a dilemma about what to choose. However, RF would be the winner in most of the cases. Here are some justifications:

- GBTs train one tree at a time, but RF can train multiple trees in parallel. So the training time is lower with RF. However, in some special cases, training and using a smaller number of trees with GBTs is faster and more convenient.
- RFs are less prone to overfitting. In other words, RFs reduces variance with more trees, but GBTs reduce bias with more trees.
- RFs can be easier to tune since performance improves monotonically with the number of trees, but GBTs perform badly with an increased number of trees.

Random forest for classification

We're familiar with the customer churn prediction problem from Chapter 3, *Scala for Learning Classification*, and we also know the data well. We also know the working principle of RF. So, we can directly jump into coding using the Spark-based implementation of RFs.

We get started by instantiating a `RandomForestClassifier` estimator by invoking the `RandomForestClassifier()` interface:

```
val rf = new RandomForestClassifier()
                    .setLabelCol("label")
                    .setFeaturesCol("features")
                    .setSeed(1234567L)  // for reproducibility
```

Now that we have three transformers and an estimator ready, the next task is to chain in a single pipeline, that is, each of them acts as a stage:

```
val pipeline = new Pipeline()
        .setStages(Array(PipelineConstruction.ipindexer,
                PipelineConstruction.labelindexer,
                    PipelineConstruction.assembler,rf))
```

Let's define `paramGrid` to perform a grid search over the hyperparameter space:

```
val paramGrid = new ParamGridBuilder()
            .addGrid(rf.maxDepth, 3 :: 5 :: 15 :: 20 :: 50 :: Nil)
            .addGrid(rf.featureSubsetStrategy, "auto" :: "all" :: Nil)
            .addGrid(rf.impurity, "gini" :: "entropy" :: Nil)
            .addGrid(rf.maxBins, 2 :: 5 :: 10 :: Nil)
            .addGrid(rf.numTrees, 10 :: 50 :: 100 :: Nil)
            .build()
```

Let's define a `BinaryClassificationEvaluator` evaluator to evaluate the model:

```
val evaluator = new BinaryClassificationEvaluator()
                    .setLabelCol("label")
                    .setRawPredictionCol("prediction")
```

We use a `CrossValidator` to perform 10-fold cross validation to select the best model:

```
val crossval = new CrossValidator()
        .setEstimator(pipeline)
        .setEvaluator(evaluator)
        .setEstimatorParamMaps(paramGrid)
        .setNumFolds(numFolds)
```

Let's call now the `fit` method so that the complete predefined pipeline, including all feature preprocessing and the DT classifier, is executed multiple times—each time with a different hyperparameter vector:

```
val cvModel = crossval.fit(Preprocessing.trainDF)
```

Now it's time to evaluate the predictive power of the DT model on the test dataset.

As a first step, we need to transform the test set with the model pipeline, which will map the features according to the same mechanism we described in the feature engineering step:

```
val predictions = cvModel.transform(Preprocessing.testSet)
prediction.show(10)
```

This will lead us to the following DataFrame showing the predicted labels against the actual labels. Additionally, it shows the raw probabilities:

```
+-----+---------------+--------------------+
|label|Predicted_label|         probability|
+-----+---------------+--------------------+
|    0|            1.0|          [0.0,1.0]|
|    0|            0.0|          [1.0,0.0]|
|    0|            0.0|          [1.0,0.0]|
|    0|            0.0|          [1.0,0.0]|
|    1|            1.0|          [0.0,1.0]|
|    1|            1.0|          [0.0,1.0]|
|    1|            1.0|          [0.0,1.0]|
|    1|            1.0|          [0.0,1.0]|
|    1|            0.0|[0.69565217391304...|
|    1|            1.0|          [0.0,1.0]|
+-----+---------------+--------------------+
only showing top 10 rows
```

However, based on the preceding prediction DataFrame, it is really difficult to guess the classification accuracy.

But in the second step, the evaluation is done using `BinaryClassificationEvaluator` as follows:

```
val accuracy = evaluator.evaluate(predictions)
println("Classification accuracy: " + accuracy)
```

The following is the output:

Accuracy: 0.8800055207949945

So we get about 87% classification accuracy from our binary classification model. Now, similar to SVM and LR, we will observe the area under the precision-recall curve and the area under the ROC curve based on the following RDD, which contains the raw scores on the test set:

```scala
val predictionAndLabels = predictions
    .select("prediction", "label")
    .rdd.map(x => (x(0).asInstanceOf[Double], x(1)
      .asInstanceOf[Double]))
```

The preceding RDD can be used to compute the previously mentioned performance metrics:

```scala
val metrics = new BinaryClassificationMetrics(predictionAndLabels)
println("Area under the precision-recall curve: " + metrics.areaUnderPR)
println("Area under the receiver operating characteristic (ROC) curve: " +
metrics.areaUnderROC)
```

In this case, the evaluation returns 88% accuracy but only 73% precision, which is much better than SVM and LR:

```
Area under the precision-recall curve: 0.7321042166486744
Area under the receiver operating characteristic (ROC) curve:
0.8800055207949945
```

Then we calculate some more metrics, for example, false and true positive and negative predictions, which will be useful to evaluate the model's performance:

```scala
val TC = predDF.ccount() //Total count

val tp = tVSpDF.filter($"prediction" === 0.0).filter($"label" ===
$"prediction")
              .count() / TC.toDouble // True positive rate
val tn = tVSpDF.filter($"prediction" === 1.0).filter($"label" ===
$"prediction")
              .count() / TC.toDouble // True negative rate
val fp = tVSpDF.filter($"prediction" === 1.0).filter(not($"label" ===
$"prediction"))
              .count() / TC.toDouble // False positive rate
val fn = tVSpDF.filter($"prediction" === 0.0).filter(not($"label" ===
$"prediction"))
              .count() / TC.toDouble // False negative rate
```

Additionally, we compute the Matthews correlation coefficient:

```scala
val MCC = (tp * tn - fp * fn) / math.sqrt((tp + fp) * (tp + fn) * (fp + tn)
* (tn + fn))
```

Let's observe how high the model confidence is:

```
println("True positive rate: " + tp *100 + "%")
println("False positive rate: " + fp * 100 + "%")
println("True negative rate: " + tn * 100 + "%")
println("False negative rate: " + fn * 100 + "%")
println("Matthews correlation coefficient: " + MCC)
```

Now let's take a look at the true positive, false positive, true negative, and false negative rates. Additionally, we see the MCC:

```
True positive rate: 0.7691154422788605
False positive rate: 0.08845577211394302
True negative rate: 0.12293853073463268
False negative rate: 0.019490254872563718
Matthews correlation coefficient: 0.6505449208932913
```

Just like DT and GBT, RF not only shows robust performance but also a slightly improved performance. And like DT and GBT, RF can be debugged to get the DT that was constructed during the classification. For the tree to be printed and the most important features selected, try the last few lines of code in the DT, and you're done.

 Can you guess how many different models were trained? Well, we have 10-folds on cross-validation and 5-dimensional hyperparameter space cardinalities between 2 and 7. Now let's do some simple math: *10 * 7 * 5 * 2 * 3 * 6 = 12,600* models!

Now that we have seen how to use RF in a classification setting, let's see another example of regression analysis.

Random forest for regression

Since RF is fast and scalable enough for a large-scale dataset, Spark-based implementations of RF help you achieve massive scalability. Fortunately, we already know the working principles of RF.

 If the proximities are calculated in RF, the storage requirements also grow exponentially.

We can jump directly into coding using the Spark-based implementation of RF for regression. We get started by instantiating a `RandomForestClassifier` estimator by invoking the `RandomForestClassifier()` interface:

```
val rfModel = new RandomForestRegressor()
       .setFeaturesCol("features")
       .setLabelCol("label")
```

Now let's create a grid space by specifying some hyperparameters, such as the max number of bins, max depth of the trees, number of trees, and impurity types:

```
// Search through decision tree's maxDepth parameter for best model
var paramGrid = new ParamGridBuilder()
       .addGrid(rfModel.impurity, "variance" :: Nil)// variance for
regression
       .addGrid(rfModel.maxBins, 25 :: 30 :: 35 :: Nil)
       .addGrid(rfModel.maxDepth, 5 :: 10 :: 15 :: Nil)
       .addGrid(rfModel.numTrees, 3 :: 5 :: 10 :: 15 :: Nil)
       .build()
```

For a better and more stable performance, let's prepare the k-fold cross-validation and grid search as part of the model tuning. As you can guess, I am going to perform 10-fold cross-validation. Feel free to adjust the number of folds based on your settings and dataset:

```
println("Preparing K-fold Cross Validation and Grid Search: Model tuning")
val numFolds = 10  // 10-fold cross-validation
val cv = new CrossValidator()
       .setEstimator(rfModel)
       .setEvaluator(new RegressionEvaluator)
       .setEstimatorParamMaps(paramGrid)
       .setNumFolds(numFolds)
```

Fantastic! We have created the cross-validation estimator. Now it's time to train the random forest regression model with cross-validation:

```
println("Training model with RandomForestRegressor algorithm")
val cvModel = cv.fit(trainingData)
```

Now that we have the fitted model, we can make predictions. Let's start evaluating the model on the train and validation sets and calculate RMSE, MSE, MAE, and R squared:

```
println("Evaluating the model on the test set and calculating the
regression metrics")
val trainPredictionsAndLabels = cvModel.transform(testData).select("label",
"prediction")
                                       .map { case Row(label: Double,
prediction: Double)
```

```
                                        => (label, prediction) }.rdd

val testRegressionMetrics = new
RegressionMetrics(trainPredictionsAndLabels)
```

Once we have the best-fitted and cross-validated model, we can expect a good prediction accuracy. Now let's observe the result on the train and validation sets:

```
val results =
"\n======================================================================\n"
+
      s"TrainingData count: ${trainingData.count}\n" +
      s"TestData count: ${testData.count}\n" +
"======================================================================\n" +
      s"TestData MSE = ${testRegressionMetrics.meanSquaredError}\n" +
      s"TestData RMSE = ${testRegressionMetrics.rootMeanSquaredError}\n" +
      s"TestData R-squared = ${testRegressionMetrics.r2}\n" +
      s"TestData MAE = ${testRegressionMetrics.meanAbsoluteError}\n" +
      s"TestData explained variance =
${testRegressionMetrics.explainedVariance}\n" +
"======================================================================\n"
println(results)
```

The following output shows the MSE, RMSE, R-squared, MAE and explained variance on the test set:

```
======================================================================
TrainingData count: 80
TestData count: 55
======================================================================
TestData MSE = 5.99847335425882
TestData RMSE = 2.4491780977011084
TestData R-squared = 0.4223425609926217
TestData MAE = 2.0564380367107646
TestData explained variance = 20.340666319995183
======================================================================
```

Great! We have managed to compute the raw prediction on the train and the test sets, and we can see the improvements compared to the LR, DT, and GBT regression models. Let's hunt for the model that helps us to achieve better accuracy:

```
val bestModel = cvModel.bestModel.asInstanceOf[RandomForestRegressionModel]
```

Additionally, we can see how the decisions were made by seeing DTs in the forest:

```
println("Decision tree from best cross-validated model: " +
bestModel.toDebugString)
```

In the following output, the `toDebugString()` method prints the tree's decision nodes and final prediction outcomes at the end leaves:

```
Decision tree from best cross-validated model with 10 trees
   Tree 0 (weight 1.0):
     If (feature 0 <= 16.0)
      If (feature 2 <= 1.0)
       If (feature 15 <= 0.0)
        If (feature 13 <= 0.0)
         If (feature 16 <= 0.0)
          If (feature 0 <= 3.0)
           If (feature 3 <= 0.0)
            Predict: 6.128571428571427
           Else (feature 3 > 0.0)
            Predict: 3.3999999999999986
 . . . .
   Tree 9 (weight 1.0):
     If (feature 0 <= 22.0)
      If (feature 2 <= 1.0)
       If (feature 1 <= 1.0)
        If (feature 0 <= 1.0)
         Predict: 3.4
 . . .
```

With RF, it is possible to measure the feature importance so that at a later stage, we can decide which features to use and which ones to drop from the DataFrame. Let's find the feature importance out of the best model we just created before we arrange all the feature in an ascending order as follows:

```
val featureImportances = bestModel.featureImportances.toArray

val FI_to_List_sorted = featureImportances.toList.sorted.toArray
println("Feature importance generated by the best model: ")
for(x <- FI_to_List_sorted) println(x)
```

Following is the feature importance generated by the model:

```
Feature importance generated by the best model:
 0.0
 0.0
 5.767724652714395E-4
 0.001616872851121874
 0.006381209526062637
 0.008867810069950395
 0.009420668763121653
 0.01802097742361489
 0.026755738338777407
 0.02761531441902482
```

```
0.031208534172407782
0.033620224027091
0.03801721834820778
0.05263475066123412
0.05562565266841311
0.13221209076999635
0.5574261654957049
```

The last result is important for understanding the feature importance. As seen, some features have higher weights than others. Even some of these have zero weights. Higher the weights the higher the importance of a feature. For example, the last two features are the most important, and the first two are less important. We can drop some unimportant columns and train the RF model to observe whether there is any reduction in the R-squared and MAE values on the test set.

What's next?

So far, we have mostly covered classic and tree-based algorithms for both regression and classification. We saw that the ensemble technique showed the best performance compared to classic algorithms. However, there are other algorithms, such as one-vs-rest algorithm, which work for solving classification problems using other classifiers, such as logistic regression.

Apart from this, neural-network-based approaches, such as **multilayer perceptron** (**MLP**), **convolutional neural network** (**CNN**), and **recurrent neural network** (**RNN**), can also be used to solve supervised learning problems. However, as expected, these algorithms require a large number of training samples and a large computing infrastructure. The datasets we used so far throughout the examples had a few samples. Moreover, those were not so high dimensional. This doesn't mean that we cannot use them to solve these two problems; we can, but this results in huge overfitting due to a lack of training samples.

How do we fix this issue? Well, we can either search for other datasets or generate training data randomly. We'll discuss and show how we can train neural-network-based deep learning models to solve other problems.

Summary

In this chapter, we had a brief introduction to powerful tree-based algorithms, such as DTs, GBT, and RF, for solving both classification and regression tasks. We saw how to develop these classifiers and regressors using tree-based and ensemble techniques. Through two real-world classification and regression problems, we saw how tree ensemble techniques outperform DT-based classifiers or regressors.

We covered supervised learning for both classification and regression on structured and labeled data. However, with the rise of cloud computing, IoT, and social media, unstructured data is growing unprecedentedly, giving more than 80% data, most of which is unlabeled.

Unsupervised learning techniques, such as clustering analysis and dimensionality reduction, are key applications in data-driven research and industry settings to find hidden structures from unstructured datasets automatically. There are many clustering algorithms, such as k-means and bisecting k-means. However, these algorithms cannot perform well with high-dimensional input datasets and often suffer from the *curse of dimensionality*. Reducing the dimensionality using algorithms such as **principal component analysis** (**PCA**) and feeding the latent data is useful for clustering billions of data points.

In the next chapter, we will use one kind of genomic data to cluster a population according to their predominant ancestry, also called geographic ethnicity. We will also learn how to evaluate the clustering analysis result and about the dimensionality reduction technique to avoid the curse of dimensionality.

5
Scala for Dimensionality Reduction and Clustering

In the previous chapters, we saw several examples of supervised learning, covering both classification and regression. We performed supervised learning techniques on structured and labelled data. However, as we mentioned previously, with the rise of cloud computing, IoT, and social media, unstructured data is increasing unprecedentedly. Collectively, more than 80% of this data is unstructured and which most of them are unlabeled.

Unsupervised learning techniques, such as clustering analysis and dimensionality reduction, are two of the key applications in data-driven research and industry settings for finding hidden structures in unstructured datasets. There are many clustering algorithms being proposed for this, such as k-means, bisecting k-means, and the Gaussian mixture model. However, these algorithms cannot perform with high-dimensional input datasets and often suffer from the *curse of dimensionality*. So, reducing the dimensionality using algorithms like **principal component analysis** (**PCA**) and feeding the latent data is a useful technique for clustering billions of data points.

In this chapter, we will use a genetic variant (one kind of genomic data) to cluster the population according to their predominant ancestry, also called geographic ethnicity. We will evaluate the clustering analysis result, followed by the dimensionality reduction technique, to avoid the curse of dimensionality.

We will cover the following topics in this chapter:

- Overview of unsupervised learning
- Learning clustering—clustering geographic ethnicity
- Dimensionality reduction with PCA
- Clustering with reduced dimensional data

Technical requirements

Make sure Scala 2.11.x and Java 1.8.x are installed and configured on your machine.

The code files of this chapters can be found on GitHub:

```
https://github.com/PacktPublishing/Machine-Learning-with-Scala-Quick-Start-
Guide/tree/master/Chapter05
```

Check out the following video to see the Code in Action:
```
http://bit.ly/2ISwb3o
```

Overview of unsupervised learning

In unsupervised learning, an input set is provided to the system during the training phase. In contrast to supervised learning, the input objects are not labeled with their class. Although in classification analysis the training dataset is labeled, we do not always have that advantage when we collect data in the real world, but still we want to find important values or hidden structures of the data. In NeuralIPS' 2016, Facebook AI Chief Yann LeCun introduced the *cake analogy*:

> *"If intelligence was a cake, unsupervised learning would be the cake, supervised learning would be the icing on the cake, and reinforcement learning would be the cherry on the cake. We know how to make the icing and the cherry, but we don't know how to make the cake."*

In order to create such a cake, several unsupervised learning tasks, including clustering, dimensionality reduction, anomaly detection, and association rule mining, are used. If unsupervised learning algorithms help find previously unknown patterns in a dataset without needing a label, we can learn the following analogy for this chapter:

- K-means is a popular clustering analysis algorithm for grouping similar data points together
- A dimensionality reduction algorithm, such as PCA, helps find the most relevant features in a dataset

In this chapter, we'll discuss these two techniques for cluster analysis with a practical example.

Clustering analysis

Clustering analysis and dimensionality reduction are the two most popular examples of unsupervised learning, which we will discuss throughout this chapter with examples. Suppose you have a large collection of legal MP3 files in your computer or smartphones. In such a case, how could you possibly group songs together if you do not have direct access to their metadata?

One possible approach could be to mix various ML techniques, but clustering is often the best solution. This is because we can develop a clustering model in order to automatically group similar songs and organize them into your favorite categories, such as country, rap, or rock.

Although the data points are not labeled, we can still do the necessary feature engineering and group similar objects together, which is commonly referred to as clustering.

 A cluster refers to a collection of data points grouped together based on certain similarity measures.

However, this is not easy for a human. Instead, a standard approach is to define a similarity measure between two objects and then look for any cluster of objects that is more similar to each other than it is to the objects in the other clusters. Once we have done the clustering of the data points (that is, the MP3 files) and the validation is completed, we know the pattern of the data (that is, what type of MP3 files fall in which group).

The left-hand side diagram shows all the **MP3 tracks in a playlist**, which are scattered. The right-hand side part shows how the MP3 are clustered based on genre:

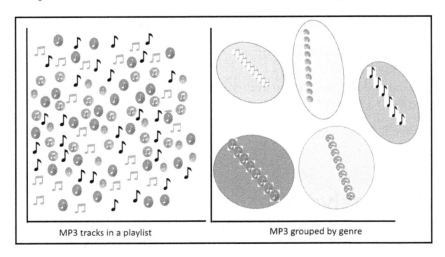

MP3 tracks in a playlist   MP3 grouped by genre

Clustering analysis algorithms

The goal of a clustering algorithm is to group a similar set of unlabeled data points together to discover underlying patterns. Here are some of the algorithms that have been proposed and used for clustering analysis:

- K-means
- Bisecting k-means
- **Gaussian mixture model (GMM)**
- **Power iteration clustering (PIC)**
- **Latent Dirichlet Allocation (LDA)**
- Streaming k-means

K-means, bisecting k-means, and GMM are the most widely used. They will be covered in detail to show a quick-start clustering analysis. However, we will also look at an example based on only k-means.

K-means for clustering analysis

K-means looks for a fixed number k of clusters (that is, the number of centroids), partitions the data points into k clusters, and allocates every data point to the nearest cluster by keeping the centroids as small as possible.

 A centroid is an imaginary or real location that represents the center of the cluster.

K-means computes the distance (usually the Euclidean distance) between data points and the center of the k clusters by minimizing the cost function, called **within-cluster sum of squares (WCSS)**. The k-means algorithm proceeds by alternating between two steps:

- **Cluster assignment step**: Each data point is assigned to the cluster whose mean has the least-squared Euclidean distance, yielding the lowest WCSS
- **Centroid update step**: The new means of the observations in the new clusters are calculated and used as the new centroids

The preceding steps can be represented in the following diagram:

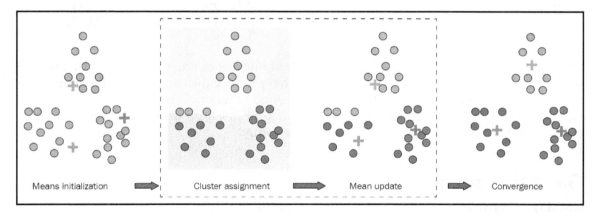

The k-means algorithm is done when the centroids have stabilized or when the predefined number of iterations have been iterated. Although k-means uses Euclidean distance, there are other ways to calculate the distance too, for example:

- The **Chebyshev distance** can be used to measure the distance by considering only the most notable dimensions

- The **Hamming distance** algorithm can identify the difference between two strings
- To make the distance metric scale-undeviating, **Mahalanobis distance** can be used to normalize the covariance matrix
- The **Manhattan distance** is used to measure the distance by considering only axis-aligned directions
- The **Minkowski distance** algorithm is used to make the Euclidean distance, Manhattan distance, and Chebyshev distance
- The **haversine distance** is used to measure the great-circle distances between two points on a sphere from the location, that is, longitudes and latitudes

Bisecting k-means

Bisecting k-means can be thought of as a combination of k-means and hierarchical clustering, which starts with all the data points in a single cluster. Then, it randomly picks a cluster to split, which returns two sub-clusters using basic k-means. This is called the **bisecting step**.

The bisecting k-means algorithm is based on a paper titled *A Comparison of Document Clustering Techniques* by *Michael Steinbach et al., KDD Workshop on Text Mining, 2000*, which has been extended to fit with Spark MLlib.

Then, the bisecting step is iterated for a predefined number of times (usually set by the user/developer), and all the splits are collected that produce the cluster with the highest similarity. These steps are continued until the desired number of clusters is reached. Although bisecting k-means is faster than regular k-means, it produces different clustering because bisecting k-means initializes clusters randomly.

Gaussian mixture model

GMM is a probabilistic model with a strong assumption that all the data points are generated from a mixture of a finite number of Gaussian distributions with unknown parameters. So, it is a distribution-based clustering algorithm too, which is based on an expectation-maximization approach.

GMM can also be considered as a generalized k-means where the model parameters are optimized iteratively to fit the model better to the training dataset. The overall process can be written in a three-step pseudocode:

- **Objective function**: Compute and maximize the log-likelihood using **expectation-maximization (EM)**
- **EM step**: This EM step consists of two sub-steps called expectation and maximization:
  - **Step E**: Compute the posterior probability of the nearer data points
  - **Step M**: Update and optimize the model parameters for fitting mixture-of-Gaussian models
- **Assignment**: Perform soft assignment during *step E*

The preceding steps can be visualized very naively as follows:

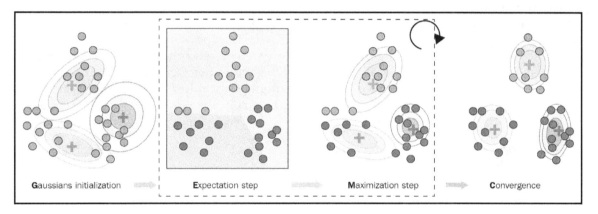

Gaussians initialization Expectation step Maximization step Convergence

Other clustering analysis algorithms

The other clustering algorithms includes PIC, which is used to cluster the nodes of a graph based on the given pairwise similarities, such as edge. The LDA is used often in text clustering use cases, such as topic modeling.

On the other hand, streaming k-means is similar to k-means but applicable for streaming data. For example, when we want to estimate the clusters dynamically so that the clustering assignment will be updated when new data arrives, using streaming k-means is a good option. For a more detailed discussion with examples, interested readers can refer to the following links:

- Spark ML-based clustering algorithms (`https://spark.apache.org/docs/latest/ml-clustering.html`)
- Spark MLlib-based clustering algorithms (`https://spark.apache.org/docs/latest/mllib-clustering.html`)

Clustering analysis through examples

One of the most important tasks in clustering analysis is the analysis of genomic profiles to attribute individuals to specific ethnic populations, or the analysis of nucleotide haplotypes for diseases susceptibility. Human ancestry from Asia, Europe, Africa, and the Americas can be separated based on their genomic data. Research has shown that the Y chromosome lineage can be geographically localized, forming the evidence for clustering the human alleles of the human genotypes. According to National Cancer Institute (`https://www.cancer.gov/publications/dictionaries/genetics-dictionary/def/genetic-variant`):

> *"Genetic variants are an alteration in the most common DNA nucleotide sequence. The term variant can be used to describe an alteration that may be benign, pathogenic, or of unknown significance. The term variant is increasingly being used in place of the term mutation."*

A better understanding of genetic variations assists us in finding correlating population groups, identifying patients who are predisposed to common diseases, and solving rare diseases. In short, the idea is to cluster geographic ethnic groups based on their genetic variants. However, before going into this any further, let's get to know the data.

Description of the dataset

The data from the 1,000 Genomes Project is a large catalog of human genetic variants. This project is meant for determining genetic variants with frequencies above 1% in the populations that were studied. The third phase of the 1,000 Genomes Project finished in September 2014, covering 2,504 individuals from 26 populations and 84,4000,000 million genetic variants. The population samples are grouped into five super-population groups, according to their predominant ancestry:

- East Asian (CHB, JPT, CHS, CDX, and KHV)
- European (CEU, TSI, FIN, GBR, and IBS)
- African (YRI, LWK, GWD, MSL, ESN, ASW, and ACB)
- American (MXL, PUR, CLM, and PEL)
- South Asian (GIH, PJL, BEB, STU, and ITU)

Each genotype comprises 23 chromosomes and a separate panel file that contains sample and population information. The data in **Variant Call Format** (**VCF**) as well the panel file can be downloaded from `ftp://ftp.1000genomes.ebi.ac.uk/vol1/ftp/release/20130502/`.

Preparing the programming environment

Since the third release of the 1,000 Genome Project contributes about 820 GB of data, using scalable software and hardware is required to process them. To do so, we will use a software stack that consists of the following components:

- **ADAM**: This can be used to achieve the scalable genomics-data-analytics platform with support for the VCF file format so that we can transform genotype-based RDD into Spark DataFrames.
- **Sparkling Water**: H20 is an AI platform for machine learning and a web-based data-processing UI with support for programming languages such as Java, Python, and R. In short, Sparkling Water equals H2O plus Spark.
- Spark-ML-based k-means is trained for clustering analysis.

For this example, we need to use multiple technology and software stacks, such as Spark, H2O, and Adam. Before using H20, make sure that your laptop has at least 16 GB RAM and sufficient storage space. I will develop this solution as a Maven project.

Let's define the properties tag on the `pom.xml` file for the Maven-friendly project:

```
<properties>
        <spark.version>2.4.0</spark.version>
        <scala.version>2.11.7</scala.version>
        <h2o.version>3.22.1.1</h2o.version>
        <sparklingwater.version>2.4.1</sparklingwater.version>
        <adam.version>0.23.0</adam.version>
</properties>
```

Once you create a Maven project on Eclipse (from an IDE or using the `mvn install` command), all the required dependencies will be downloaded!

Clustering geographic ethnicity

24 VCF files contribute around 820 GB of data, which will impose a great computational challenge. To overcome this, use the genetic variants from the smallest chromosome, Y. The size of the VCF file for this is around 160 MB. Let's get started by creating `SparkSession`:

```
val spark:SparkSession = SparkSession
            .builder()
             .appName("PopStrat")
             .master("local[*]")
              .config("spark.sql.warehouse.dir", "temp/")
               .getOrCreate()
```

Now, let's show Spark the path of both VCF and the PANEL file:

```
val genotypeFile =
"Downloads/ALL.chr22.phase3_shapeit2_mvncall_integrated_v5a.20130502.genoty
pes.vcf"
val panelFile = "Downloads/integrated_call_samples_v3.20130502.ALL.panel"
```

We process the PANEL file using Spark to access the target population data and identify the population groups. First, we create a set of `populations` that we want to form clusters:

```
val populations = Set("FIN", "GBR", "ASW", "CHB", "CLM")
```

Then, we need to create a map between the sample ID and the given population so that we can filter out the samples we are not interested in:

```
def extract(file: String, filter: (String, String) => Boolean): Map[String,
String] = {
        Source
          .fromFile(file)
          .getLines()
```

```
      .map(line => {
        val tokens = line.split(Array('\t', ' ')).toList
        tokens(0) -> tokens(1)
      })
      .toMap
      .filter(tuple => filter(tuple._1, tuple._2))
    }

val panel: Map[String, String] = extract(
      panelFile,(sampleID: String, pop: String) =>
populations.contains(pop))
```

The panel file produces the sample ID of all individuals, population groups, ethnicity, super-population groups, and genders:

Sample ID	Pop Group	Ethnicity	Super pop. group	Gender
HG00096	GBR	British in England and Scotland	EUR	male
HG00171	FIN	Finnish in Finland	EUR	female
HG00472	CHS	Southern Han Chinese	EAS	male
HG00551	PUR	Puerto Ricans from Puerto Rico	AMR	female

Check out the details of the panel file at `ftp://ftp.1000genomes.ebi.ac.uk/vol1/ftp/release/20130502/integrated_call_samples_v3.20130502.ALL.panel`.

Then, load the ADAM genotypes and filter the genotypes so that we are left with only those in the populations we are interested in:

```
val allGenotypes: RDD[Genotype] = sc.loadGenotypes(genotypeFile).rdd
val genotypes: RDD[Genotype] = allGenotypes.filter(genotype => {
      panel.contains(genotype.getSampleId)
    })
```

Next, convert the `Genotype` objects into our own `SampleVariant` objects to conserve memory. Then, the `genotype` object is converted into a `SampleVariant` object that contains the data that needs to be processed further:

- **Sample ID**: To uniquely identify a particular sample
- **Variant ID**: To uniquely identify a particular genetic variant
- **Alternate alleles count**: Needed when the sample differs from the reference genome

The signature that prepares a `SampleVariant` is given as follows, which takes `sampleID`, `variationId`, and the `alternateCount` objects:

```
// Convert the Genotype objects to our own SampleVariant objects to try and
conserve memory
case class SampleVariant(sampleId: String, variantId: Int, alternateCount:
Int)
```

Then, we have to find the `variantID` from the genotype file. A `varitantId` is a String type that consists of a name, and the start and end positions in the chromosome:

```
def variantId(genotype: Genotype): String = {
    val name = genotype.getVariant.getContigName
    val start = genotype.getVariant.getStart
    val end = genotype.getVariant.getEnd
    s"$name:$start:$end"
}
```

Once we have the `variantID`, we should hunt for the `alternateCount`. In the genotype file, objects for which an allele reference is present, would be the genetic alternates:

```
def alternateCount(genotype: Genotype): Int = {
    genotype.getAlleles.asScala.count(_ != GenotypeAllele.REF)
}
```

Finally, we will construct a `SampleVariant` object. For this, we need to intern sample IDs as they will be repeated a lot in a VCF file:

```
def toVariant(genotype: Genotype): SampleVariant = {
    new SampleVariant(genotype.getSampleId.intern(),
      variantId(genotype).hashCode(),
      alternateCount(genotype))
}
```

Now, we need is to prepare `variantsRDD`. First, we have to group the variants by sample ID so that we can process the variants sample by sample. Then, we can get the total number of samples to be used to find the variants that are missing for some samples. Finally, we have to group the variants by variant ID and filter out those variants that are missing from some samples:

```
val variantsRDD: RDD[SampleVariant] = genotypes.map(toVariant)
val variantsBySampleId: RDD[(String, Iterable[SampleVariant])] =
variantsRDD.groupBy(_.sampleId)
val sampleCount: Long = variantsBySampleId.count()

println("Found " + sampleCount + " samples")
val variantsByVariantId: RDD[(Int, Iterable[SampleVariant])] =
```

```
variantsRDD.groupBy(_.variantId).filter {
  case (_, sampleVariants) => sampleVariants.size == sampleCount
}
```

Now, let's map `variantId` with the count of samples with an alternate count of greater than zero. Then, we filter out the variants that are not in our desired frequency range. The objective here is to reduce the number of dimensions in the dataset to make it easier to train the model:

```
val variantFrequencies: collection.Map[Int, Int] = variantsByVariantId
  .map {
    case (variantId, sampleVariants) =>
      (variantId, sampleVariants.count(_.alternateCount > 0))
  }
  .collectAsMap()
```

The total number of samples (or individuals) has been determined. Now, before grouping them using their variant IDs, we can filter out less significant variants. Since we have more than 84 million genetic variants, filtering would help us deal with the curse of dimensionality.

The specified range is arbitrary and was chosen because it includes a reasonable number of variants, but not too many. To be more specific, for each variant, the frequency for alternate alleles has been calculated, and variants with fewer than 12 alternate alleles have been excluded, leaving about 3,000,000 variants in the analysis (for 23 chromosome files):

```
val permittedRange = inclusive(11, 11) // variants with less than 12
alternate alleles
val filteredVariantsBySampleId: RDD[(String, Iterable[SampleVariant])] =
  variantsBySampleId.map {
    case (sampleId, sampleVariants) =>
      val filteredSampleVariants = sampleVariants.filter(
        variant =>
          permittedRange.contains(
            variantFrequencies.getOrElse(variant.variantId, -1)))
      (sampleId, filteredSampleVariants)
  }
```

Once we have `filteredVariantsBySampleId`, we need to sort the variants for each sample ID. Each sample should now have the same number of sorted variants:

```
val sortedVariantsBySampleId: RDD[(String, Array[SampleVariant])] =
    filteredVariantsBySampleId.map {
      case (sampleId, variants) =>
        (sampleId, variants.toArray.sortBy(_.variantId))
    }
  println(s"Sorted by Sample ID RDD: " +
sortedVariantsBySampleId.first())
```

All items in the RDD should now have the same variants in the same order. The final task is to use `sortedVariantsBySampleId` to construct an RDD of a row that contains the region and the alternate count:

```
val rowRDD: RDD[Row] = sortedVariantsBySampleId.map {
    case (sampleId, sortedVariants) =>
      val region: Array[String] = Array(panel.getOrElse(sampleId,
"Unknown"))
      val alternateCounts: Array[Int] =
sortedVariants.map(_.alternateCount)
      Row.fromSeq(region ++ alternateCounts)
  }
```

Therefore, we can just use the first one to construct our header for the training DataFrame:

```
val header = StructType(
    Array(StructField("Region", StringType)) ++
      sortedVariantsBySampleId
      .first()
      ._2
      .map(variant => {
        StructField(variant.variantId.toString, IntegerType)
      }))
```

Well done! We have our RDD and the `StructType` header. Now, we can play with the Spark machine learning algorithm with minimal adjustment/conversion.

Training the k-means algorithm

Once we have the `rowRDD` and the header, we need to construct the rows of our schema DataFrame from the variants using the header and `rowRDD`:

```
// Create the SchemaRDD from the header and rows and convert the SchemaRDD
into a Spark DataFrame
val sqlContext = sparkSession.sqlContext
```

```
var schemaDF = sqlContext.createDataFrame(rowRDD, header)
schemaDF.show(10)
>>>
```

The preceding `show()` method should show a snapshot of the training dataset that contains the features and the `label` columns (that is, Region):

```
+------+-----------+-----------+-----------+-----------+-----------+-----------+
|Region|-2099651974|-2086354790|-2051555302|-2033308294|-2031516326|-1948345126|
+------+-----------+-----------+-----------+-----------+-----------+-----------+
|   GBR|          1|          1|          0|          0|          0|          0|
|   CHB|          0|          0|          0|          0|          0|          0|
|   CHB|          0|          0|          0|          0|          0|          0|
|   ASW|          0|          0|          0|          0|          0|          0|
|   ASW|          0|          0|          0|          0|          0|          0|
|   CHB|          0|          0|          0|          0|          0|          0|
|   ASW|          0|          0|          0|          0|          0|          1|
|   GBR|          0|          0|          0|          0|          0|          0|
|   ASW|          0|          0|          0|          0|          0|          0|
|   ASW|          0|          0|          0|          0|          0|          1|
+------+-----------+-----------+-----------+-----------+-----------+-----------+
only showing top 10 rows
```

In the preceding DataFrame, only a few `feature` columns and the `label` column are shown so that it fits on the page. Since the training would be unsupervised, we need to drop the `label` column (that is, Region):

```
schemaDF = sqlContext.createDataFrame(rowRDD, header).drop("Region")
schemaDF.show(10)
>>>
```

The preceding `show()` method shows the following snapshot of the training dataset for k-means. Note that there is no `label` column (that is, Region):

```
+-----------+-----------+-----------+-----------+-----------+-----------+-----------+-----------+
|-2099651974|-2086354790|-2051555302|-2033308294|-2031516326|-1948345126|-1933334022|-1900872614|
+-----------+-----------+-----------+-----------+-----------+-----------+-----------+-----------+
|          1|          1|          0|          0|          0|          0|          0|          0|
|          0|          0|          0|          0|          0|          0|          0|          0|
|          0|          0|          0|          0|          0|          0|          0|          0|
|          0|          0|          0|          0|          0|          0|          0|          0|
|          0|          0|          0|          0|          0|          0|          0|          0|
|          0|          0|          0|          0|          0|          1|          0|          0|
|          0|          0|          0|          0|          0|          0|          0|          0|
|          0|          0|          0|          0|          0|          0|          1|          0|
|          0|          0|          0|          0|          0|          1|          1|          0|
+-----------+-----------+-----------+-----------+-----------+-----------+-----------+-----------+
```

In Chapter 1, *Introduction to Machine Learning with Scala,* and Chapter 2, *Scala for Regression Analysis,* we saw that Spark expects two columns (features and label) for supervised training. However, for the unsupervised training, only a single column containing the features is required. Since we dropped the label column, we now need to amalgamate the entire variable column into a single features column. For this, we will use the VectorAssembler() transformer. Let's select the columns to be embedded into a vector space:

```
val featureCols = schemaDF.columns
```

Then, we will instantiate the VectorAssembler() transformer by specifying the input columns and the output column:

```
// Using vector assembler to create feature vector
val featureCols = schemaDF.columns
val assembler = new VectorAssembler()
    .setInputCols(featureCols)
    .setOutputCol("features")

val assembleDF = assembler.transform(schemaDF).select("features")
```

Now, let's see what the feature vectors for the k-means look like:

```
assembleDF.show()
```

The preceding line shows the assembled vectors, which can be used as the feature vectors for the k-means model:

```
+--------------------+
|            features|
+--------------------+
|(59,[0,1],[1.0,1.0])|
|(59,[35,51],[1.0,...|
|(59,[39,42],[1.0,...|
|(59,[11,16,18,28,...|
|(59,[9,16],[1.0,1...|
|(59,[28,35],[1.0,...|
|(59,[5,9,10,16,28...|
|    (59,[17],[1.0])|
|(59,[6,9,13,21,44...|
|(59,[5,6,18,19,21...|
|    (59,[31],[1.0])|
|(59,[6,11,51],[1....|
|(59,[13,38,39],[1...|
|(59,[11,17,28,35,...|
|    (59,[35],[1.0])|
|    (59,[12],[1.0])|
|     (59,[2],[1.0])|
|(59,[6,10,38,43,4...|
|(59,[0,1,29,34,51...|
|(59,[21,26],[1.0,...|
+--------------------+
only showing top 20 rows
```

Finally, we are ready to train the k-means algorithm and evaluate the clustering by computing **WCSS**:

```
val kmeans = new KMeans().setK(5).setSeed(12345L)
val model = kmeans.fit(assembleDF)

val WCSS = model.computeCost(assembleDF)
println("Within Set Sum of Squared Errors for k = 5 is " + WCSS)
  }
```

The following is the **WCSS** value for k = 5:

```
Within Set Sum of Squared Errors for k = 5 is 59.34564329865
```

We managed to apply k-means to cluster genetic variants. However, we saw that the WCSS was high because k-means was unable to separate the non-linearity among different correlated and high-dimensional features. This is because genomic sequencing datasets are very high dimensional due to a huge number of genetic variants.

In the next section, we will see how we can use dimensionality-reduction techniques, such as PCA, to reduce the dimensionality of the input data before feeding it to k-means in order to get better clustering quality.

Dimensionality reduction

Since humans are visual creatures, understanding a high dimensional dataset (even with more than three dimensions) is impossible. Even for a machine (or say, our machine learning algorithm), it's difficult to model the non-linearity from correlated and high-dimensional features. Here, the dimensionality reduction technique is a savior.

Statistically, dimensionality reduction is the process of reducing the number of random variables to find a low-dimensional representation of the data while preserving as much information as possible.

The overall step in PCA can be visualized naively in the following diagram:

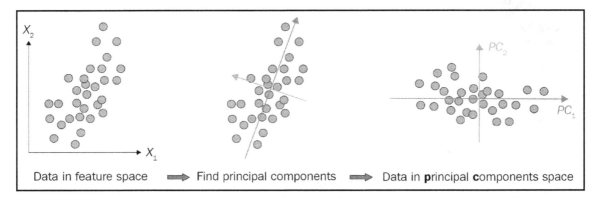

PCA and **singular-value decomposition (SVD)** are the most popular algorithms for dimensionality reduction. Technically, PCA is a statistical technique that's used to emphasize variation and extract the most significant patterns (that is, features) from a dataset, which is not only useful for clustering but also for classification and visualization.

Principal component analysis with Spark ML

Spark-ML-based PCA can be used to project vectors to a low-dimensional space to reduce the dimensionality of genetic variant features before feeding them into the k-means model. The following example shows how to project 6D feature vectors into 4D principal components from the following feature vector:

```scala
val data = Array(
      Vectors.dense(1.2, 3.57, 6.8, 4.5, 2.25, 3.4),
      Vectors.dense(4.60, 4.10, 9.0, 5.0, 1.67, 4.75),
      Vectors.dense(5.60, 6.75, 1.11, 4.5, 2.25, 6.80))

val df = spark.createDataFrame(data.map(Tuple1.apply)).toDF("features")
df.show(false)
```

Now that we have a feature DataFrame with a 6D-feature vector, it can be fed into the PCA model:

```
+---------------------------+
|features                   |
+---------------------------+
|[1.2,3.57,6.8,4.5,2.25,3.4] |
|[4.6,4.1,9.0,5.0,1.67,4.75] |
|[5.6,6.75,1.11,4.5,2.25,6.8]||
+---------------------------+
```

First, we have to instantiate the PCA model by setting the necessary parameters, as follows:

```
val pca = new PCA()
        .setInputCol("features")
        .setOutputCol("pcaFeatures")
        .setK(4)
        .fit(df)
```

To distinguish the original features from the principal component-based features, we set the output column name as pcaFeatures using the setOutputCol() method. Then, we set the dimension of the PCA (that is, the number of principal components). Finally, we fit the DataFrame to make the transformation. A model can be loaded from older data but will have an empty vector for explainedVariance. Now, let's show the resulting features:

```
val result = pca.transform(df).select("features", "pcaFeatures")
result.show(false)
```

The preceding code produces a feature DataFrame with 4D feature vectors as principal components using the PCA:

```
+----------------------------------------------------------------------+
|pcaFeatures                                                           |
+----------------------------------------------------------------------+
|[2.8828586569325583,5.9776156708087615,-6.2640619020647,1.3642364212641955]|
|[3.007898962240862,10.345013628637705,-6.264061902064701,1.364236421264195]|
|[-5.388955893878964,8.210457814128072,-6.264061902064701,1.364236421264196]|
+----------------------------------------------------------------------+
```

Similarly, we can transform the assembled DataFrame (that is, assembleDF) in the previous step and the top five principle components. You can adjust the number of principal components, though.

Finally, to avoid any ambiguity, we renamed the `pcaFeatures` column to `features`:

```
val pcaDF = pca.transform(assembleDF)
          .select("pcaFeatures")
          .withColumnRenamed("pcaFeatures", "features")
pcaDF.show()
```

The preceding lines of code show the embedded vectors, which can be used as the feature vectors for the k-means model:

```
+--------------------+
|            features|
+--------------------+
|[-0.0322320102812...|
|[0.04745047735089...|
|[-0.0332140726994...|
|[-0.1110921441100...|
|[-0.0375778701056...|
|[-0.0340456681672...|
|[-0.1083958692025...|
|[-0.0177665347420...|
|[0.05380817444571...|
|[-0.1014376410370...|
|[-0.0178548747978...|
|[0.08328541553353...|
|[0.02980402336806...|
|[-0.1038847392578...|
|[-0.0155973063597...|
|[0.01140843693081...|
|[-0.0158289900287...|
|[0.03367559270881...|
|[-0.0024716378274...|
|[-0.0346824191993...|
+--------------------+
only showing top 20 rows
```

The preceding screenshot shows the top five principal components as the most important features. Excellent—everything went smoothly. Finally, we are ready to train the k-means algorithm and evaluate clustering by computing WCSS:

```
val kmeans = new KMeans().setK(5).setSeed(12345L)
val model = kmeans.fit(pcaDF)

val WCSS = model.computeCost(pcaDF)
println("Within Set Sum of Squared Errors for k = 5 is " + WCSS)
    }
```

This time, the WCSS is slightly lower (compared to the previous value, which was `59.34564329865`):

```
Within Set Sum of Squared Errors for k = 5 is 52.712937492025276
```

Normally, we set the number of k (that is, 5) randomly and computed the WCSS. However, this way, we cannot always set the optimal number of clusters. In order to find an optimal value, researchers have come up with two techniques, called the elbow method and silhouette analysis, which we'll look at in the following subsection.

Determining the optimal number of clusters

Sometimes, assuming the number of clusters naively and before starting the training may not be a good idea. If the assumption is too far from the optimal number of clusters, the model performs poorly because of the overfitting or underfitting issue that's introduced. So, determining the number of optimal clusters is a separate optimization problem. There are two popular techniques to tackle this:

- The heuristic approach, called the **elbow method**
- **Silhouette analysis**, to observe the separation distance between predicted clusters

The elbow method

We start by setting the k value to 2 and running the k-means algorithm on the same dataset by increasing k and observing the value of WCSS. As expected, a drastic drop should occur in the cost function (that is, WCSS values) at some point. However, after the drastic fall, the value of WCSS becomes marginal with the increasing value of k. As suggested by the elbow method, we can pick the optimal value of k after the last big drop of WCSS:

```
val iterations = 20
    for (k <- 2 to iterations) {
      // Trains a k-means model.
      val kmeans = new KMeans().setK(k).setSeed(12345L)
      val model = kmeans.fit(pcaDF)

// Evaluate clustering by computing Within Set Sum of Squared Errors.
val WCSS = model.computeCost(pcaDF)
println("Within Set Sum of Squared Errors for k = " + k + " is " + WCSS)
    }
```

Now, let's see the WCSS values for a different number of clusters, such as between 2 and 20:

```
Within Set Sum of Squared Errors for k = 2 is 135.0048361804504
Within Set Sum of Squared Errors for k = 3 is 90.95271589232344
. . .
Within Set Sum of Squared Errors for k = 19 is 11.505990055606803
Within Set Sum of Squared Errors for k = 20 is 12.26634441065655
```

As shown in the preceding code, we calculated the cost function, WCSS, as a function of a number of clusters for the k-means algorithm, and applied them to the Y chromosome genetic variants from the selected population groups. It can be observed that a big drop occurs when k = 5 (which is not a drastic drop, though). Therefore, we chose a number of clusters to be 10.

The silhouette analysis

Analyzing the silhouette is carried out by observing the separation distance between predicted clusters. Drawing a silhouette plot will show the distance between a data point from its neighboring clusters, and then we can visually inspect a number of clusters so that similar data points get well-separated.

The silhouette score, which is used to measure the clustering quality, has a range of [-1, 1]. Evaluate the clustering quality by computing the silhouette score:

```
val evaluator = new ClusteringEvaluator()
for (k <- 2 to 20 by 1) {
    val kmeans = new KMeans().setK(k).setSeed(12345L)
    val model = kmeans.fit(pcaDF)
    val transformedDF = model.transform(pcaDF)
    val score = evaluator.evaluate(transformedDF)
    println("Silhouette with squared Euclidean distance for k = " + k + "
is " + score)
    }
```

We get the following output:

```
Silhouette with squared Euclidean distance for k = 2 is 0.9175803927739566
Silhouette with squared Euclidean distance for k = 3 is 0.8288633816548874
. . . .
Silhouette with squared Euclidean distance for k = 19 is 0.5327466913746908
Silhouette with squared Euclidean distance for k = 20 is
0.45336547054142284
```

As shown in the preceding code, the height value of the silhouette is generated with k = 2, which is 0.9175803927739566. However, this suggests that genetic variants should be clustered in two groups. The elbow method suggested k = 5 as the optimal number of clusters.

Let's find out the silhouette using the squared Euclidean distance, as shown in the following code block:

```
val kmeansOptimal = new KMeans().setK(2).setSeed(12345L)
val modelOptimal = kmeansOptimal.fit(pcaDF)

// Making predictions
val predictionsOptimalDF = modelOptimal.transform(pcaDF)
predictionsOptimalDF.show()

// Evaluate clustering by computing Silhouette score
val evaluatorOptimal = new ClusteringEvaluator()
val silhouette = evaluatorOptimal.evaluate(predictionsOptimalDF)
println(s"Silhouette with squared Euclidean distance = $silhouette")
```

The silhouette with the squared Euclidean distance for k = 2 is 0.9175803927739566.

It has been found that the bisecting k-means algorithm can result in better cluster assignment for data points, converging to the global minima. On the other hand, k-means often gets stuck in the local minima. Please note that you might observe different values of the preceding parameters depending on your machine's hardware configuration and the random nature of the dataset.

 Interested readers should also refer to the Spark-MLlib-based clustering techniques at https://spark.apache.org/docs/latest/mllib-clustering.html for more insights.

Summary

In this chapter, we discussed some clustering analysis techniques, such as k-means, bisecting k-means, and GMM. We saw a step-by-step example of how to cluster ethnic groups based on their genetic variants. In particular, we used the PCA for dimensionality reduction, k-means for clustering, and H2O and ADAM for handling large-scale genomics datasets. Finally, we learned about the elbow and silhouette methods for finding the optimal number of clusters.

Clustering is the key to most data-driven applications. Readers can try to apply clustering algorithms on higher-dimensional datasets, such as gene expression or miRNA expression, in order to cluster similar and correlated genes. A great resource is the gene expression cancer RNA-Seq dataset, which is open source. This dataset can be downloaded from the UCI machine learning repository at `https://archive.ics.uci.edu/ml/datasets/gene+expression+cancer+RNA-Seq`.

In the next chapter, we will discuss item-based collaborative filtering approaches for the recommender system. We'll learn how to develop a book recommendation system. Technically, it will be a model-based recommendation engine with Scala and Spark. We will see how we can interoperate between ALS and matrix factorization.

6
Scala for Recommender System

In this chapter, we will learn about different approaches for developing recommender systems. Then we will learn how to develop a book recommendation system. Technically, it will be a model-based recommendation engine based on **alternating least squares** (**ALS**) and matrix factorization algorithms. We will use Spark MLlib-based implementation of these algorithms in Scala. In a nutshell, we will learn the following topics throughout this chapter:

- Overview of recommendation systems
- Similarity-based recommender system
- Content-based recommender system
- Collaborative approaches
- Hybrid recommendation systems
- Developing a model-based book recommendation system

Technical requirements

Make sure Scala 2.11.x and Java 1.8.x are installed and configured on your machine.

The code files of this chapters can be found on GitHub:

```
https://github.com/PacktPublishing/Machine-Learning-with-Scala-Quick-Start-
Guide/tree/master/Chapter06
```

Check out the following video to see the Code in Action:
`http://bit.ly/2UQTFHs`

Overview of recommendation systems

A recommender system is an information filtering approach, which predicts the rating given by a user to an item. Then the item for which the predicted rating is high will be recommended to the user. Recommender systems are now being used more or less everywhere for recommending movies, music, news, books, research articles, products, videos, books, news, Facebook friends, restaurants, routes, search queries, social tags, products, collaborators, jokes, restaurants, garments, financial services, Twitter pages, Android/iOS apps, hotels, life insurance, and even partners, in online dating sites.

Types of recommender systems

There are a couple of ways to develop recommendation engines that typically produce a list of recommendations, such as similarity-based, content-based, collaborative, and hybrid recommendation systems as shown in the following figure:

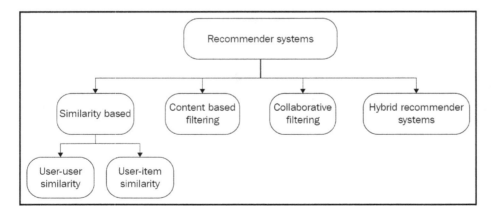

We will discuss the similarity-based, content-based, collaborative, and hybrid recommendation systems. Then based on their pros and cons, we will see a hands-on example showing how to develop a book recommendation system.

Similarity-based recommender systems

There are two main types of similarity-based approaches: **user-user similarity** and **user-item similarity**. These can be used to build recommendation systems. To use a user-user item similarity approach, first construct a user-user similarity matrix. It will then pick items that are already liked by similar users and, finally, it recommends items for a specific user.

Suppose we want to develop a book recommender system: naturally, there will be many book users (readers) and a list of books. For the sake of brevity, let's pick the following machine learning-related books as the representative ones for the readers:

Then a user-user similarity based recommender system will recommend books based on a similarity measure using some similarity measure techniques. For example, the cosine similarity is calculated as follows:

$$Sim(A, B) = cos(\theta) = \frac{A.B}{|A||B|}$$

In the preceding equation, *A* and *B* represent two users. If the similarity threshold is greater than or equal to a defined threshold, users *A* and *B* will most likely have similar preferences:

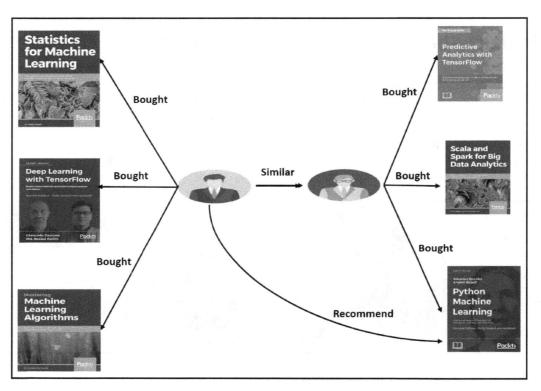

However, user-user similarity based recommender systems are not robust. There are several reasons for that:

- User preferences and tastes usually change over time
- They are computationally very expensive because of the similarity calculation for so many cases from very sparse matrix calculation

Amazon and YouTube have millions of subscribed users, so any user-user utility matrix that you created would be a very sparse one. One workaround is using item-item similarity, which also computes an item-item utility matrix, finding similar items and, finally, recommending similar items, just like in the following diagram:

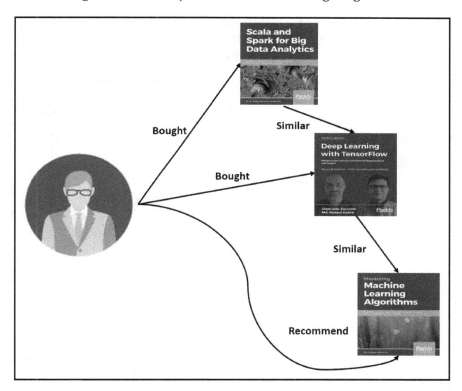

This approach has one advantage over the user-user similarity approach, which is that usually the ratings on a given item do not change very significantly after an initial period. Let's take as an example the book *The Hundred Page Machine Learning Book*, which has already got a very good rating on Amazon even though it was released just a few months ago. So, even if over the next few months a few people give it lower ratings, its ratings would not change much after the initial period.

Interestingly, this is also an assumption that the ratings will not change very significantly over time. However, this assumption works very well in cases where the number of users is much higher than the number of items.

Content-based filtering approaches

Content-based filtering approaches are based on classical machine learning techniques such as classification or regression. This type of system learns how to represent an item (book) I_j and a user U_i. Then, a separate feature matrix for both I_j and U_i are created before combining them as a feature vector. Then the feature vector is fed into a classification or regression model for the training. This way, the ML model generates the label L_{ij}, which is interestingly the corresponding rating given by the user U_i on the item I_j:

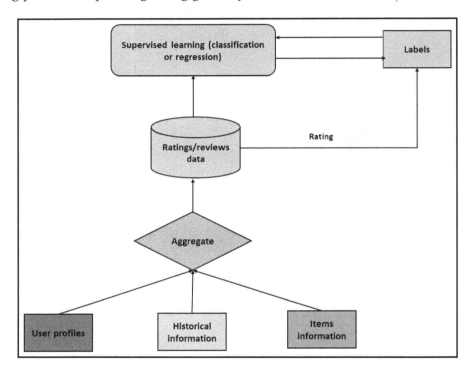

A general warning is that the features should be created so they have direct impact on the rating (**Labels**). This means features should be as dependent as possible to avoid correlations.

Collaborative filtering approaches

The idea of collaborative filtering is that when we have many users who liked some items, then those items can be recommended to users who have not seen them yet. Suppose we have four readers and four books, as shown in the following diagram:

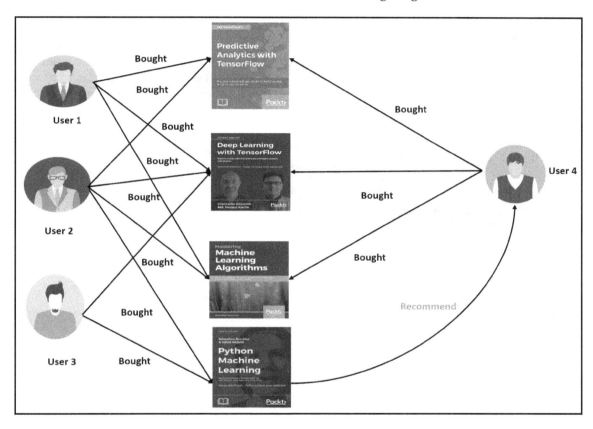

Also, imagine all of these users have bought item 1 (that is, **Predictive Analytics with TensorFlow**) and item 2 (that is, **Deep Learning with TensorFlow**). Now, suppose **User 4** has read items 1, 2, and 3 and say both **User 1** and **User 2** have bought item 3 (that is, **Mastering Machine Learning Algorithms**). However, since **User 4** has not seen item 4 (that is, **Python Machine Learning**) yet, **User 3** can recommend it to him.

So, the basic assumption is that users who have recommended an item previously tend to give recommendations in the future, too. If this assumption does not hold any longer, then a collaborative filtering recommender system cannot be build. This is probably the reason collaborative filtering approaches suffer from cold start, scalability, and sparsity problems.

Cold start: Collaborative filtering approaches can get stuck and cannot make recommendation especially when a large amount of data about users is missing in the uer-item matrix.

The utility matrix

Suppose we have a group of users who show a preference for a set of books. The higher a user's preference for a book, the higher the rating would be, between 1 and 10. Let's try to understand the problem using a matrix, with rows representing users and columns representing books:

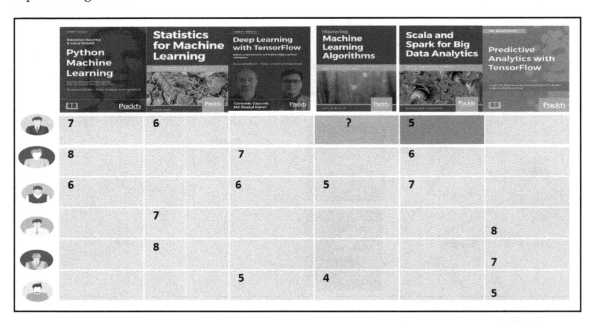

	Python Machine Learning	Statistics for Machine Learning	Deep Learning with TensorFlow	Machine Learning Algorithms	Scala and Spark for Big Data Analytics	Predictive Analytics with TensorFlow
	7	6		?	5	
	8		7		6	
	6		6	5	7	
		7				8
		8				7
			5	4		5

Let's assume that ratings range from 1 to 10, with 10 being the highest level of preference. Then, in the preceding table, a user (row 1) gives a rating of **7** for the first book (column 1) and rates the second book as a **6**. Also, there are many empty cells that indicate users have not given any ratings for those books.

This matrix is often called a user-item or utility matrix, where each row represents a user and each column represents an item (book), while a cell represents the corresponding rating given by the user to that item.

In practice, the utility matrix is *very sparse* because a large number of cells are empty. The reason is that we have so many items and it is almost impossible for a single user to give ratings to all of the items. Even if a user rates 10% of the items, the other 90% of the cells of this matrix will still be empty. These empty cells often represented by NaN, which means not a number, although in our example utility matrix we used **?**. This sparsity often creates computational complexity. Let me give you an example.

Suppose there are 1 million users (*n*) and only 10,000 items (movies, *m*), which is *10,000,000 * 10,000* or 10^{11}, a very large number. Now, even if a user has rated 10 books, this means that the total number of given ratings will be *10 * 1 million = 10^7*. The sparsity of this matrix can be calculated as follows:

$$S_m = Number\ of\ empty\ cells\ /\ Total\ number\ of\ cells = (10^{10} - 10^7)/10^{10} = 0.9999$$

This means 99.99% of the cells will still be empty.

Model-based book recommendation system

In this section, we will show how to develop a model-based book recommendation system with the Spark MLlib library. Books and the corresponding ratings were downloaded from this link: `http://www2.informatik.uni-freiburg.de/~cziegler/BX/`. There are three CSV files:

- `BX-Users.csv`: Contains user's demographic data and each user is specified with user IDs (`User-ID`).

- `BX-Books.csv`: Book related information such as `Book-Title`, `Book-Author`, `Year-Of-Publication`, and `Publisher` are there. Each book is identified by an ISBN. Also, `Image-URL-S`, `Image-URL-M`, and `Image-URL-L` are given.

- `BX-Book-Ratings.csv`: Contains the rating specified by the `Book-Rating` column. Ratings are on a scale from 1 to 10 (higher values denoting higher appreciation), or implicit, expressed by 0.

Before we jump into the coding part, we need to know a bit more about the matrix factorization techniques such as **singular value decomposition** (**SVD**). SVD can be used to transform both the item and the user entries into the same potential space, which represents the interaction between users and items. The rationale behind matrix decomposition is that potential features represent how users score items.

Matrix factorization

So, given the description of the users and the items, the task here is to predict how the user will rate those items that have not yet been rated. More formally, if a user U_i likes item V_1, V_5, and V_7, then the task is to recommend item V_j to user U_i that they will most probably like too as shown in the following figure:

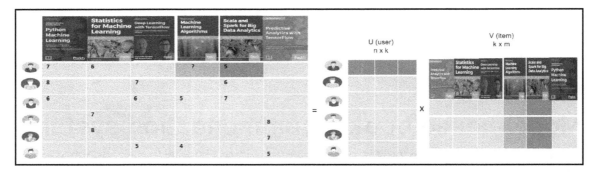

Once we have such an application, the idea is that each time we receive new data, we update it to the training dataset and then update the model obtained by ALS training, where the collaborative filtering method is used. To handle the user-book utility matrix, a low-rank matrix factorization algorithm is used:

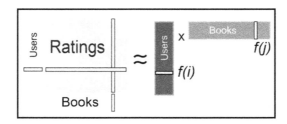

Since not all the books are rated by all the users, not all of the entries in this matrix are known. The collaborative filtering approach discussed in a preceding section comes to this party as the savior. Well, using collaborative filtering, we can solve an optimization problem to approximate the ratings matrix by factorizing **User factors (V)** and **Book factors (V)**, which can be depicted as follows:

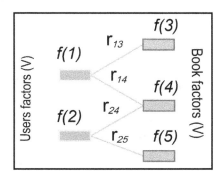

These two matrices are selected such that the error for the users-book pairs (in the case of known rating) gets minimized. The ALS algorithm first fills the user matrix with random values (between 1 and 10, in our case) and then optimizes those values such that the error is minimized. Then the ALS holds the book matrix as fixed and optimizes the value of the user's matrix using the following mathematical equation:

$$f[i] = \operatorname*{arg\,min}_{w \in \mathfrak{R}^d} \sum_{j \in N_{brs}(i)} (r_{ij} - w^T f[j])^2 + \lambda ||w||_2^2$$

Spark MLlib supports a model-based collaborative filtering approach. In such an approach, users and items are described by a small set of latent factors for predicting missing entries of a user-item utility matrix. As described earlier, the ALS algorithms can learn those latent factors in an iterative way. The ALS algorithm accepts six parameters, namely `numBlocks`, `rank`, `iterations`, `lambda`, `implicitPrefs`, and `alpha`. `numBlocks` is number of blocks required to parallelize the computation. The `rank` parameter is the number of latent factors. The `iterations` parameter is the number of iterations by which ALS will get converged. The `lambda` parameter signifies the regularization parameter. The `implicitPrefs` parameter means that we want to use explicit feedback from the other users, and, finally, `alpha` is the baseline confidence in preference observations.

Exploratory analysis

In this subsection, we will perform some exploratory analysis about the ratings, books, and related statistics. This analysis will help us understand the data well:

```
val ratigsFile = "data/BX-Book-Ratings.csv"
var ratingDF = spark.read.format("com.databricks.spark.csv")
    .option("delimiter", ";")
    .option("header", true)
    .load(ratigsFile)
```

The following code segments show you the DataFrame of books from the BX-Books.csv file:

```
/* Explore and query on books        */
val booksFile = "data/BX-Books.csv"
var bookDF = spark.read.format("com.databricks.spark.csv")
            .option("header", "true")
            .option("delimiter", ";")
            .load(booksFile)
bookDF = bookDF.select(bookDF.col("ISBN"),
                    bookDF.col("Book-Title"),
                    bookDF.col("Book-Author"),
                    bookDF.col("Year-Of-Publication"))

bookDF = bookDF.withColumnRenamed("Book-Title", "Title")
            .withColumnRenamed("Book-Author", "Author")
            .withColumnRenamed("Year-Of-Publication", "Year")
bookDF.show(10)
```

The following is the output:

```
+----------+--------------------+--------------------+----+
|      ISBN|               Title|              Author|Year|
+----------+--------------------+--------------------+----+
|0195153448|  Classical Mythology|  Mark P. O. Morford|2002|
|0002005018|         Clara Callan|Richard Bruce Wright|2001|
|0060973129|Decision in Normandy|        Carlo D'Este|1991|
|0374157065|Flu: The Story of...|    Gina Bari Kolata|1999|
|0393045218|The Mummies of Ur...|    E. J. W. Barber|1999|
|0399135782|The Kitchen God's...|            Amy Tan|1991|
|0425176428|What If?: The Wor...|      Robert Cowley|2000|
|0671870432|       PLEADING GUILTY|        Scott Turow|1993|
|0679425608|Under the Black F...|    David Cordingly|1996|
|074322678X|Where You'll Find...|        Ann Beattie|2002|
+----------+--------------------+--------------------+----+
only showing top 10 rows
```

Let's see how many distinct books there are:

```
val numDistinctBook = bookDF.select(bookDF.col("ISBN")).distinct().count()
println("Got " + numDistinctBook + " books")
```

The following is the output:

Got 271,379 books

This information will be valuable for a later case, so that we can know how many books are missing ratings in the rating dataset. To register both datasets, we can use the following code:

```
ratingsDF.createOrReplaceTempView("ratings")
moviesDF.createOrReplaceTempView("books")
```

This will help to make the in-memory querying faster by creating a temporary view as a in-memory table. Let's check the ratings-related statistics. Just use the following code lines:

```
/* Explore and query ratings for books         */
val numRatings = ratingDF.count()
val numUsers = ratingDF.select(ratingDF.col("UserID")).distinct().count()
val numBooks = ratingDF.select(ratingDF.col("ISBN")).distinct().count()
println("Got " + numRatings + " ratings from " + numUsers + " users on " +
numBooks + " books")
```

You should find Got 1149780 ratings from 105283 users on 340556 books. Now, let's get the maximum and minimum ratings along with the count of users who have rated a book:

```
// Get the max, min ratings along with the count of users who have rated a
book.
val statDF = spark.sql("select books.Title, bookrates.maxRating,
bookrates.minRating, bookrates.readerID "
        + "from(SELECT ratings.ISBN,max(ratings.Rating) as maxRating,"
        + "min(ratings.Rating) as minRating,count(distinct UserID) as
readerID "
        + "FROM ratings group by ratings.ISBN) bookrates "
        + "join books on bookrates.ISBN=books.ISBN " + "order by
bookrates.readerID desc")

    statDF.show(10)
```

The preceding code should generate the max and min ratings, along with the count of users who have rated a book:

```
+--------------------+---------+---------+--------+
|               Title|maxRating|minRating|readerID|
+--------------------+---------+---------+--------+
|         Wild Animus|        9|        0|    2502||
|The Lovely Bones:...|        9|        0|    1295|
|    The Da Vinci Code|       9|        0|     883|
|Divine Secrets of...|        9|        0|     732|
|The Red Tent (Bes...|        9|        0|     723|
|      A Painted House|       9|        0|     647|
|The Secret Life o...|        9|        0|     615|
|Snow Falling on C...|        9|        0|     614|
|  Angels &amp; Demons|       9|        0|     586|
|Where the Heart I...|        9|        0|     585|
+--------------------+---------+---------+--------+
only showing top 10 rows
```

Now, to get further insight we need to know more about the users and their ratings, which can be done by finding the top ten most active users and how many times they have rated a book:

```
// Show the top 10 most-active users and how many times they rated a book
val mostActiveReaders = spark.sql("SELECT ratings.UserID, count(*) as CT
from ratings "
    + "group by ratings.UserID order by CT desc limit 10")
mostActiveReaders.show()
```

The preceding lines of code should show the top ten most active users and how many times they have rated a book:

```
+------+-----+
|UserID|   CT|
+------+-----+
| 11676|13602|
|198711| 7550|
|153662| 6109|
| 98391| 5891|
| 35859| 5850|
|212898| 4785|
|278418| 4533|
| 76352| 3367|
|110973| 3100|
|235105| 3067|
+------+-----+
```

Now let's have a look at a particular user, and find the books that, say, user `130554` rated higher than `5`:

```
// Find the movies that user 130554 rated higher than 5
val ratingBySpecificReader = spark.sql(
      "SELECT ratings.UserID, ratings.ISBN,"
      + "ratings.Rating, books.Title FROM ratings JOIN books "
      + "ON books.ISBN=ratings.ISBN "
      + "WHERE ratings.UserID=130554 and ratings.Rating > 5")

ratingBySpecificReader.show(false)
```

As described, the preceding line of code should show the name of all the movies rated by user 130554 giving more than 5 ratings:

```
+------+----------+------+--------------------+
|UserID|      ISBN|Rating|               Title|
+------+----------+------+--------------------+
|130554|0152273220|     6|Black Horses for ...|
|130554|0671794019|     8|Jennifer Murdley'...|
|130554|0375822062|     7|George's Marvelou...|
|130554|0394864042|     7|The Mystery of th...|
|130554|0590877356|     6| Survival (Remnants)|
|130554|0786817194|    10|Monster Manor: Vo...|
|130554|0439405572|    10|Guardians Of Ga'h...|
|130554|0440400651|     7|Grumpy Pumpkins (...|
|130554|0671023098|     8|SHARK BITE: AGAIN...|
|130554|0679889167|     6|Squire (Protector...|
|130554|0689704550|     6|The Mushroom Cent...|
|130554|0807580872|     6|Tree House Myster...|
|130554|068815526X|     6|Behold...the Drag...|
|130554|0440412897|     9|   The Chocolate Touch|
|130554|0590603795|     6|The Berenstain Be...|
|130554|0590997262|     6|The Stranger (Ani...|
|130554|0786819146|    10|The Eternity Code...|
|130554|080755510X|     6|The Mystery of th...|
|130554|0679810455|     7|Robin Hood (Bulls...|
|130554|0316119202|     8|    The Enormous Egg|
+------+----------+------+--------------------+
only showing top 20 rows
```

Prepare training and test rating data

The following code splits ratings RDD into training data RDD (60%) and test data RDD (40%). The second parameter (that is 1357L) is the *seed*, which is typically used for the purpose of reproducibility:

```
val splits = ratingDF.randomSplit(Array(0.60, 0.40), 1357L)
val (trainingData, testData) = (splits(0), splits(1))

trainingData.cache
testData.cache

val numTrainingSample = trainingData.count()
val numTestSample = testData.count()
println("Training: " + numTrainingSample + " test: " + numTestSample)
```

You will see that there are 689,144 ratings in the training DataFrame and 345,774 ratings in the test DataFrame. The ALS algorithm requires an RDD of ratings for the training. The following code illustrates the way to build the recommendation model using APIs:

```
val trainRatingsRDD = trainingData.rdd.map(row => {
    val userID = row.getString(0)
    val ISBN = row.getInt(1)
    val ratings = row.getString(2)
    Rating(userID.toInt, ISBN, ratings.toDouble)
})
```

trainRatingsRDD is an RDD of ratings that contains UserID, ISBN, and the corresponding ratings from the training dataset that we prepared in the preceding step. Similarly, we prepared another RDD from the test DataFrame:

```
val testRatingsRDD = testData.rdd.map(row => {
    val userID = row.getString(0)
    val ISBN = row.getInt(1)
    val ratings = row.getString(2)
    Rating(userID.toInt, ISBN, ratings.toDouble)
})
```

Based on the trainRatingsRDD, we build an ALS user model by adding the maximal iteration, a number of blocks, alpha, rank, lambda, seed, and implicit preferences. This method is generally used for analyzing and predicting missing ratings of specific users:

```
val model : MatrixFactorizationModel = new ALS()
        .setIterations(10)
        .setBlocks(-1)
        .setAlpha(1.0)
        .setLambda(0.01)
```

```
.setRank(25)
.setSeed(1234579L)
.setImplicitPrefs(false) // We want explicit feedback
.run(trainRatingsRDD)
```

Finally, we iterated the model for learning 10 times. With this setting, we got good prediction accuracy. Readers are recommended to apply hyperparameter tuning to find the optimum values for these parameters. In order to evaluate the quality of the model, we compute the **root mean squared error (RMSE)**. The following code calculates the RMSE value for the model that was developed with the help of the training set:

```
var rmseTest = computeRmse(model, testRatingsRDD, true)
println("Test RMSE: = " + rmseTest) //Less is better
```

For the preceding setting, we get the following output:

```
Test RMSE: = 1.6867585251053991
```

The preceding method computes the RMSE to evaluate the model. The lower the RMSE, the better the model and its prediction capability is, which goes as follows:

```
//Compute the RMSE to evaluate the model. Less the RMSE better the model
and it's prediction capability.
def computeRmse(model: MatrixFactorizationModel, ratingRDD: RDD[Rating],
implicitPrefs: Boolean): Double =            {
    val predRatingRDD: RDD[Rating] = model.predict(ratingRDD.map(entry =>
(entry.user, entry.product)))
    val predictionsAndRatings = predRatingRDD.map {entry => ((entry.user,
entry.product), entry.rating)}
                                .join(ratingRDD
                                .map(entry => ((entry.user, entry.product),
entry.rating)))
                                .values
    math.sqrt(predictionsAndRatings.map(x => (x._1 - x._2) * (x._1 -
x._2)).mean()) // return MSE
            }
```

Finally, let's do some movie recommendations for a specific user. Let's get the top ten book predictions for user 276747:

```
println("Recommendations: (ISBN, Rating)")
println("---------------------------------")
val recommendationsUser = model.recommendProducts(276747, 10)
recommendationsUser.map(rating => (rating.product,
rating.rating)).foreach(println)
println("---------------------------------")
```

We get the following output:

```
Recommendations: (ISBN => Rating)
    (1051401851,15.127044702142243)
    (2056910662,15.11531283195148)
    (1013412890,14.75898119158678)
    (603241602,14.53024153450836)
    (1868529062,14.180262929540024)
    (746990712,14.121654522195225)
    (1630827789,13.741728003481194)
    (1179316963,13.571754513473993)
    (505970947,13.506755847456258)
    (632523982,13.46591014905454)
-----------------------------------
```

We believe that the performance of the preceding model could be increased more. However, as far as we know, there is no model tuning facility available for the MLlib-based ALS algorithm.

 Interested readers should refer to `https://spark.apache.org/docs/preview/ml-collaborative-filtering.html` for more on tuning the ML-based ALS models.

Adding new user ratings and making new predictions

We can create a sequence of a new user ID, the ISBN of the book, and the rating predicted in the previous step:

```
val new_user_ID = 300000 // new user ID randomly chosen

//The format of each line is (UserID, ISBN, Rating)
val new_user_ratings = Seq(
    (new_user_ID, 817930596, 15.127044702142243),
    (new_user_ID, 1149373895, 15.11531283195148),
    (new_user_ID, 1885291767, 14.75898119158678),
    (new_user_ID, 459716613, 14.53024153450836),
    (new_user_ID, 3362860, 14.180262929540024),
    (new_user_ID, 1178102612, 14.121654522195225),
    (new_user_ID, 158895996, 13.741728003481194),
    (new_user_ID, 1007741925, 13.571754513473993),
    (new_user_ID, 1033268461, 13.506755847456258),
    (new_user_ID, 651677816, 13.46591014905454))
```

```
val new_user_ratings_RDD = spark.sparkContext.parallelize(new_user_ratings)
val new_user_ratings_DF =
spark.createDataFrame(new_user_ratings_RDD).toDF("UserID", "ISBN",
"Rating")

val newRatingsRDD = new_user_ratings_DF.rdd.map(row => {
    val userId = row.getInt(0)
    val movieId = row.getInt(1)
    val ratings = row.getDouble(2)
    Rating(userId, movieId, ratings)
})
```

Now we add them to the data we will use to train our recommender model. We use Spark's
union() transformation for this:

```
val complete_data_with_new_ratings_RDD =
trainRatingsRDD.union(newRatingsRDD)
```

Finally, we train the ALS model using all the parameters we selected before (when using
the small dataset):

```
val newModel : MatrixFactorizationModel = new ALS()
        .setIterations(10)
        .setBlocks(-1)
        .setAlpha(1.0)
        .setLambda(0.01)
        .setRank(25)
        .setSeed(123457L)
        .setImplicitPrefs(false)
        .run(complete_data_with_new_ratings_RDD)
```

We will need to repeat that every time a user adds new ratings. Ideally, we will do this in
batches, and not for every single rating that comes into the system for every user. Then we
can again make recommendations for other users such as 276724, whose ratings about
books were missing previously:

```
// Making Predictions. Get the top 10 book predictions for user 276724
//Book recommendation for a specific user. Get the top 10 book predictions
for reader 276747
println("Recommendations: (ISBN, Rating)")
println("----------------------------------")
val newPredictions = newModel.recommendProducts(276747, 10)
newPredictions.map(rating => (rating.product,
rating.rating)).foreach(println)
println("----------------------------------")
```

The following is the output:

```
Recommendations: (ISBN, Rating)
---------------------------------
(1901261462,15.48152758068679)
(1992983531,14.306018295431224)
(1438448913,14.05457411015043)
(2022242154,13.516608439192192)
(817930596,13.487733919030019)
(1079754533,12.991618591680165)
(611897245,12.716161072778828)
(11041460,12.44511878072316)
(651596038,12.13345082904184)
(1955775932,11.7254312955358)
---------------------------------
```

Finally, we compute the RMSE:

```
var newrmseTest = computeRmse(newModel, testRDD, true)
println("Test RMSE: = " + newrmseTest) //Less is better
```

The following is the output:

```
Test RMSE: = 4.892434600794704
```

Summary

In this chapter, we have learned different approaches for recommender systems, such as similarity-based, content-based, collaborative filtering, and hybrid. Additionally, we discussed the downsides of these approaches. Then we implemented an end-to-end book recommendation system, which is a model-based recommendation with Spark. We have also seen how to interoperate between ALS and matrix factorization to efficiently handle a utility matrix.

In the next chapter, we will explain some basic concepts of **deep learning** (DL), which is one of the emerging branches of ML. We will briefly discuss some of the most well known and widely used neural network architectures. Then, we will look at various features of DL frameworks and libraries.

Then we will see how to prepare a programming environment, before moving on to coding with some open source DL libraries, such as **Deeplearning4j** (DL4J). Finally, we will solve a real-life problem using two neural network architectures, called **multilayer perceptron** (MLP) and **long short-term memory** (LSTM).

7
Introduction to Deep Learning with Scala

Throughout Chapter 2, *Scala for Regression Analysis*, to Chapter 6, *Scala for Recommender System*, we have learned about linear and classic **machine learning** (**ML**) algorithms through real-life examples. In this chapter, we will explain some basic concepts of **deep learning** (**DL**). We will start with DL, which is one of the emerging branches of ML. We will briefly discuss some of the most well-known and widely used neural network architectures and DL frameworks and libraries.

Finally, we will use the **Long Short-Term Memory** (**LSTM**) architecture for cancer type classification from a very high-dimensional dataset curated from **The Cancer Genome Atlas** (**TCGA**). The following topics will be covered in this chapter:

- DL versus ML
- DL and neural networks
- Deep neural network architectures
- DL frameworks
- Getting started with learning

Technical requirements

Make sure Scala 2.11.x and Java 1.8.x are installed and configured on your machine.

The code files of this chapters can be found on GitHub:

```
https://github.com/PacktPublishing/Machine-Learning-with-Scala-Quick-Start-
Guide/tree/master/Chapter07
```

Check out the following video to see the Code in Action:
`http://bit.ly/2vwrxzb`

DL versus ML

Simple ML methods that were used in small-scale data analysis are not effective anymore because the effectiveness of ML methods diminishes with large and high-dimensional datasets. Here comes DL—a branch of ML based on a set of algorithms that attempt to model high-level abstractions in data. Ian Goodfellow *et al.* (Deep Learning, MIT Press, 2016) defined DL as follows:

> *"Deep learning is a particular kind of machine learning that achieves great power and flexibility by learning to represent the world as a nested hierarchy of concepts, with each concept defined in relation to simpler concepts, and more abstract representations computed in terms of less abstract ones."*

Similar to the ML model, a DL model also takes in an input, X, and learns high-level abstractions or patterns from it to predict an output of Y. For example, based on the stock prices of the past week, a DL model can predict the stock price for the next day. When performing training on such historical stock data, a DL model tries to minimize the difference between the prediction and the actual values. This way, a DL model tries to generalize to inputs that it hasn't seen before and makes predictions on test data.

Now, you might be wondering, if an ML model can do the same tasks, why do we need DL for this? Well, DL models tend to perform well with large amounts of data, whereas old ML models stop improving after a certain point. The core concept of DL is inspired by the structure and function of the brain, which are called **artificial neural networks** (**ANNs**). Being at the core of DL, ANNs help you learn the associations between sets of inputs and outputs in order to make more robust and accurate predictions. However, DL is not only limited to ANNs; there have been many theoretical advances, software stacks, and hardware improvements that bring DL to the masses. Let's look at an example; suppose we want to develop a predictive analytics model, such as an animal recognizer, where our system has to resolve two problems:

- To classify whether an image represents a cat or a dog
- To cluster images of dogs and cats

If we solve the first problem using a typical ML method, we must define the facial features (ears, eyes, whiskers, and so on) and write a method to identify which features (typically nonlinear) are more important when classifying a particular animal.

However, at the same time, we cannot address the second problem because classical ML algorithms for clustering images (such as k-means) cannot handle nonlinear features. Take a look at the following diagram, which shows a workflow that we would follow whether we wanted to classify if the given image is of a cat:

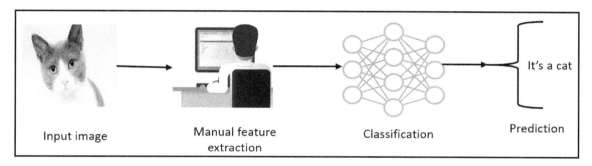

DL algorithms will take these two problems one step further, and the most important features will be extracted automatically after determining which features are the most important for classification or clustering. In contrast, when using a classical ML algorithm, we would have to provide the features manually.

A DL algorithm would take more sophisticated steps instead. For example, first, it would identify the edges that are the most relevant when clustering cats or dogs. It would then try to find various combinations of shapes and edges hierarchically. This step is called **extract, transform, and load** (**ETL**). Then after several iterations, hierarchical identification of complex concepts and features would be carried out. Then, based on the identified features, the DL algorithm would decide which of these features are most significant for classifying the animal. This step is known as feature extraction. Finally, it would take out the label column and perform unsupervised training using **autoencoders** (**AEs**) to extract the latent features to be redistributed to k-means for clustering. Then, the **clustering assignment hardening loss** (**CAH loss**) and reconstruction loss are jointly optimized toward optimal clustering assignment.

However, in practice, a DL algorithm is fed with a raw image representations, which doesn't see an image as we see it because it only knows the position of each pixel and its color. The image is divided into various layers of analysis. At a lower level, the software analyzes, for example, a grid of a few pixels with the task of detecting a type of color or various nuances. If it finds something, it informs the next level, which at this point checks whether or not that given color belongs to a larger form, such as a line.

The process continues to the upper levels until the algorithm understand what is shown in the following diagram:

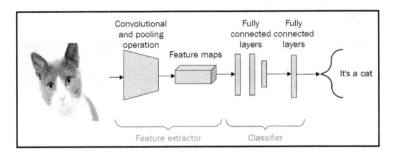

Although *dog versus cat* is an example of a very simple classifier, software that's capable of doing these types of things is now widespread and is found in systems for recognizing faces, or in those for searching an image on Google, for example. This kind of software is based on DL algorithms. On the contrary, by using a linear ML algorithm, we cannot build such applications since these algorithms are incapable of handling nonlinear image features.

Also, using ML approaches, we typically only handle a few hyperparameters. However, when neural networks are brought to the mix, things become too complex. In each layer, there are millions or even billions of hyperparameters to tune—so many that the cost function becomes non-convex. Another reason for this is that the activation functions that are used in hidden layers are nonlinear, so the cost is non-convex.

DL and ANNs

ANNs, which are inspired by how a human brain works, form the core of deep learning and its true realization. Today's revolution around deep learning would have not been possible without ANNs. Thus, to understand DL, we need to understand how neural networks work.

ANNs and the human brain

ANNs represent one aspect of the human nervous system and how the nervous system consists of a number of neurons that communicate with each other using axons. The receptors receive the stimuli either internally or from the external world. Then, they pass this information into the biological neurons for further processing.

There are a number of dendrites, in addition to another long extension called the axon. Toward its extremities, there are minuscule structures called synaptic terminals, which are used to connect one neuron to the dendrites of other neurons. Biological neurons receive short electrical impulses called signals from other neurons, and in response, they trigger their own signals.

We can thus summarize that the neuron comprises a cell body (also known as the soma), one or more dendrites for receiving signals from other neurons, and an axon for carrying out the signals that are generated by the neurons. A neuron is in an active state when it is sending signals to other neurons. However, when it is receiving signals from other neurons, it is in an inactive state. In an idle state, a neuron accumulates all the signals that are received before reaching a certain activation threshold. This whole thing motivated researchers to test out ANNs.

A brief history of neural networks

The most significant progress in ANNs and DL can be described with the following timeline. We have already seen how the artificial neurons and perceptrons provided the base in 1943 and 1958, respectively. Then, the XOR was formulated as a linearly non-separable problem in 1969 by Minsky *et al.*, but later, in 1974, Werbos *et al.* demonstrated the backpropagation algorithm for training the perceptron.

However, the most significant advancement happened in the 1980s when John Hopfield *et al.* proposed the Hopfield network in 1982. Then, one of the godfathers of the neural network and DL, Hinton and his team proposed the Boltzmann machine in 1985. However, probably one of the most significant advances happened in 1986 when Hinton *et al.* successfully trained the MLP and Jordan *et al.* proposed RNNs. In the same year, Smolensky *et al.* also proposed the improved version of Boltzmann machine called the **Restricted Boltzmann Machine (RBM)**.

However, in the 90s era, the most significant year was 1997, when Lecun *et al.* proposed LeNet in 1990, and Jordan *et al.* proposed the Recurrent Neural Network in 1997. In the same year, Schuster *et al.* proposed the improved version of LSTM and the improved version of the original RNN, called the bidirectional RNN. The following timeline provides a brief glimpse into the history of different neural network architectures:

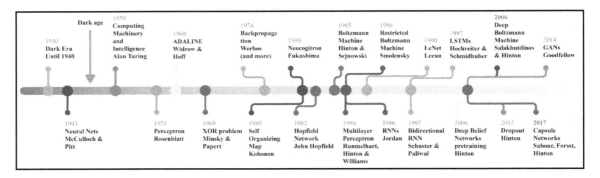

Despite significant advances in computing, from 1997 to 2005, we didn't experience much advancement until Hinton struck again in 2006 when he and his team proposed the **Deep Belief Network** (**DBN**) by stacking multiple RBMs. Then, in 2012, Hinton invented dropout, which significantly improves regularization and overfitting in deep neural networks.

After that, Ian Goodfellow *et al.* introduced GANs, which was a significant milestone in image recognition. In 2017, Hinton proposed CapsNet to overcome the limitation of regular CNNs, which was one of the most significant milestones so far.

How does an ANN learn?

Based on the concept of biological neurons, the term and idea of ANNs arose. Similar to biological neurons, the artificial neuron consists of the following:

- One or more incoming connections that aggregate signals from neurons
- One or more output connections for carrying the signal to the other neurons
- An activation function, which determines the numerical value of the output signal

Besides the state of a neuron, synaptic weight is considered, which influences the connection within the network. Each weight has a numerical value indicated by W_{ij}, which is the synaptic weight connecting neuron i to neuron j. Now, for each neuron i, an input vector can be defined by $x_i = (x_1, x_2,...x_n)$ and a weight vector can be defined by $w_i = (w_{i1}, w_{i2},...w_{in})$. Now, depending on the position of a neuron, the weights and the output function determine the behavior of an individual neuron. Then, during forward propagation, each unit in the hidden layer gets the following signal:

$$net_i = \sum_j W_{ij} X_j \ldots\ldots (a)$$

Nevertheless, among the weights, there is also a special type of weight called a bias unit, b. Technically, bias units aren't connected to any previous layer, so they don't have true activity. But still, the bias b value allows the neural network to shift the activation function to the left or right. By taking the bias unit into consideration, the modified network output is formulated as follows:

$$net_i = \sum_j W_{ij} X_j + b_j \ldots\ldots (b)$$

The preceding equation signifies that each hidden unit gets the sum of inputs, multiplied by the corresponding weight—this is known as the **Summing junction**. Then, the resultant output in the **Summing junction** is passed through the activation function, which squashes the output, as depicted in the following diagram:

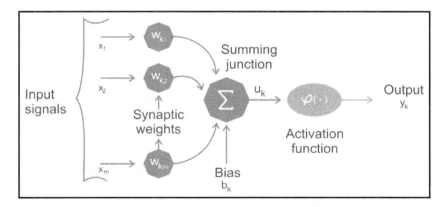

Working principal of an artificial neuron model

A practical neural network architecture, however, is composed of input, hidden, and output layers that are composed of *nodes* that make up a network structure, but still follow the working principal of an artificial neuron model, as shown in the preceding diagram. The input layer only accepts numeric data, such as features in real numbers, images with pixel values, and so on:

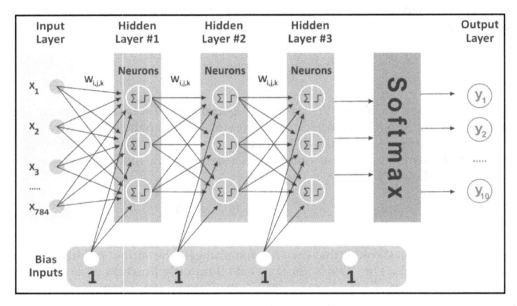

A neural network with one input layer, three hidden layers, and an output layer

Here, the hidden layers perform most of the computation to learn the patterns and the network evaluates how accurate its prediction is compared to the actual output using a special mathematical function called the loss function. It could be a complex one or a very simple mean squared error, which can be defined as follows:

$$MSE = \frac{1}{n} \sum_{i=1}^{n} \left(Y_i - \hat{Y}_i \right)^2$$

In the preceding equation, $\hat{Y}$ signifies the prediction made by the network, while Y represents the actual or expected output. Finally, when the error is no longer being reduced, the neural network converges and makes prediction through the output layer.

Training a neural network

The learning process for a neural network is configured as an iterative process of the optimization of the weights. The weights are updated in each epoch. Once the training starts, the aim is to generate predictions by minimizing the loss function. The performance of the network is then evaluated on the test set. We already know about the simple concept of an artificial neuron. However, generating only some artificial signals is not enough to learn a complex task. As such, a commonly used supervised learning algorithm is the backpropagation algorithm, which is very commonly used to train a complex ANN.

Ultimately, training such a neural network is an optimization problem too, in which we try to minimize the error by adjusting network weights and biases iteratively, by using backpropagation through **gradient descent (GD)**. This approach forces the network to backtrack through all its layers to update the weights and biases across nodes in the opposite direction of the loss function.

However, this process using GD does not guarantee that the global minimum is reached. The presence of hidden units and the nonlinearity of the output function means that the behavior of the error is very complex and has many local minimas. This backpropagation step is typically performed thousands or millions of times, using many training batches, until the model parameters converge to values that minimize the cost function. The training process ends when the error on the validation set begins to increase, because this could mark the beginning of a phase overfitting.

The downside of using GD is that it takes too long to converge, which makes it impossible to meet the demand of handling large-scale training data. Therefore, a faster GD, called **Stochastic Gradient Descent (SDG)** was proposed, which is also a widely used optimizer in DNN training. In SGD, we use only one training sample per iteration from the training set to update the network parameters, which is a stochastic approximation of the true cost gradient.

There are other advanced optimizers nowadays such as Adam, RMSProp, ADAGrad, Momentum, and so on. Each of them is either an direct or indirect optimized version of SGD.

Weight and bias initialization

Now, here's a tricky question: how do we initialize the weights? Well, if we initialize all the weights to the same value (for example, 0 or 1), each hidden neuron will get exactly the same signal. Let's try to break it down:

- If all weights are initialized to 1, then each unit gets a signal equal to the sum of the inputs
- If all weights are 0, which is even worse, then every neuron in a hidden layer will get zero signal

For network weight initialization, Xavier initialization is used widely. It is similar to random initialization but often turns out to work much better, since it can identify the rate of initialization depending on the total number of input and output neurons by default.

You may be wondering whether you can get rid of random initialization while training a regular DNN. Well, recently, some researchers have been talking about random orthogonal matrix initialization's that perform better than just any random initialization for training DNNs. When it comes to initializing the biases, we can initialize them to be zero.

But setting the biases to a small constant value, such as 0.01 for all biases, ensures that all **rectified linear units** (**ReLUs**) can propagate some gradient. However, it neither performs well nor shows consistent improvement. Therefore, sticking with zero is recommended.

Activation functions

To allow a neural network to learn complex decision boundaries, we apply a non-linear activation function to some of its layers. Commonly used functions include Tanh, ReLU, softmax, and variants of these. More technically, each neuron receives a signal of the weighted sum of the synaptic weights and the activation values of the neurons that are connected as input. One of the most widely used functions for this purpose is the so-called sigmoid logistic function, which is defined as follows:

$$Out_i = \frac{1}{(1 + e^{-x})}$$

The domain of this function includes all real numbers, and the co-domain is (0, 1). This means that any value obtained as an output from a neuron (as per the calculation of its activation state) will always be between zero and one. The **Sigmoid** function, as represented in the following diagram, provides an interpretation of the saturation rate of a neuron, from not being active (equal to **0**) to complete saturation, which occurs at a predetermined maximum value (equal to **1**):

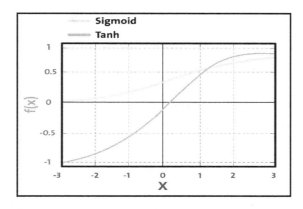

Sigmoid versus Tanh activation function

On the other hand, a hyperbolic tangent, or **Tanh**, is another form of activation function. **Tanh** flattens a real-valued number between **-1** and **1**. The preceding graph shows the difference between **Tanh** and **Sigmoid** activation functions. In particular, mathematically, *tanh* activation function can be expressed as follows:

$$tanh(x) = 2\sigma(2x) - 1$$

In general, in the last level of an **feedforward neural network** (**FFNN**), the softmax function is applied as the decision boundary. This is a common case, especially when solving a classification problem. The softmax function used for the probability distribution over the possible classes in a multiclass classification problem.

For a regression problem, we do not need to use any activation function since the network generates continuous values—that is, probabilities. However, I've seen people using the IDENTITY activation function for regression problems nowadays.

To conclude, choosing proper activation functions and network weight initializations are two problems that make a network perform at its best and help to obtain good training. Now that we know the brief history of neural networks, let's deep-dive into different architectures in the next section, which will give us an idea on their usage.

Neural network architectures

We can categorize DL architectures into four groups:

- **Deep neural networks (DNNs)**
- **Convolutional neural networks (CNNs)**
- **Recurrent neural networks (RNNs)**
- **Emergent architectures (EAs)**

However, DNNs, CNNs, and RNNs have many improved variants. Although most of the variants are proposed or developed for solving domain-specific research problems, the basic working principles still follow the original DNN, CNN, and RNN architectures. The following subsections will give you a brief introduction to these architectures.

DNNs

DNNs are neural networks that have a complex and deeper architecture with a large number of neurons in each layer, and many connections between them. Although DNN refers to a very deep network, for simplicity, we consider MLP, **stacked autoencoder (SAE)**, and **deep belief networks (DBNs)** as DNN architectures. These architectures mostly work as an FFNN, meaning information propagates from input to output layers.

Multiple perceptrons are stacked together as MLPs, where layers are connected as a directed graph. Fundamentally, an MLP is one of the most simple FFNNs since it has three layers: an input layer, a hidden layer, and an output layer. This way, the signal propagates one way, from the input layer to the hidden layers to the output layer, as shown in the following diagram:

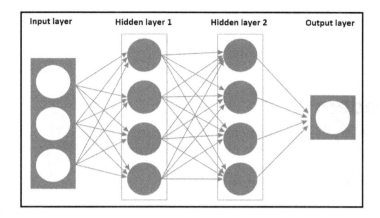

Autoencoders and RBMs are the basic building blocks for SAEs and DBNs, respectively. Unlike MLP, which is an FFNN that's trained in a supervised way, both SAEs and DBNs are trained in two phases: unsupervised pre-training and supervised fine-tuning. In unsupervised pre-training, layers are stacked in order and trained in a layer-wise manner with used, unlabeled data. In supervised fine-tuning, an output classifier layer is stacked and the complete neural network is optimized by retraining with labeled data.

One problem with MLP is that it often overfits the data, so it doesn't generalize well. To overcome this issue, DBN was proposed by Hinton *et al*. It uses a greedy, layer-by-layer, pre-training algorithm. DBNs are composed of a visible layer and multiple hidden unit layers. The building blocks of a DBN are RBMs, as shown in the following diagram, where several RBMs are stacked one after another:

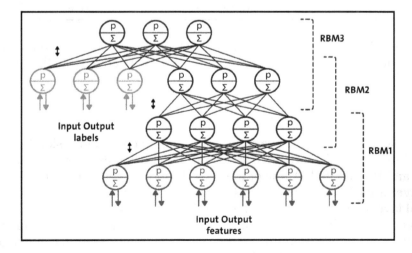

The top two layers have undirected, symmetric connections in between, but the lower layers have directed connections from the preceding layer. Despite numerous successes, DBNs are now being replaced with AEs.

Autoencoders

AEs are also special types of neural networks that learn automatically from the input data. AEs consists of two components: the encoder and the decoder. The encoder compresses the input into a latent-space representation. Then, the decoder part tries to reconstruct the original input data from this representation:

- **Encoder**: Encodes or compresses the input into a latent-space representation using a function known as $h=f(x)$
- **Decoder**: Decodes or reconstructs the input from the latent space representation using a function known as $r=g(h)$

So, an AE can be described by a function of $g(f(x)) = o$, where we want 0 as close as the original input of x. The following diagram shows how an AE typically works:

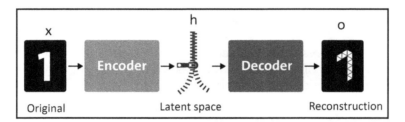

AEs are very useful at data denoising and dimensionality reduction for data visualization. AEs can learn data projections, called representations, more effectively than PCA.

CNNs

CNNs have achieved much and have a wide adoption in computer vision (for example, image recognition). In CNN networks, the connections schemes are significantly different compared to an MLP or DBN. A few of the convolutional layers are connected in a cascade style. Each layer is backed up by a ReLU layer, a pooling layer, and additional convolutional layers (+ReLU), and another pooling layer, which is followed by a fully connected layer and a softmax layer. The following diagram is a schematic of the architecture of a CNN that's used for facial recognition, which takes facial images as input and predicts emotions such as anger, disgust, fear, happy, sad and so on.

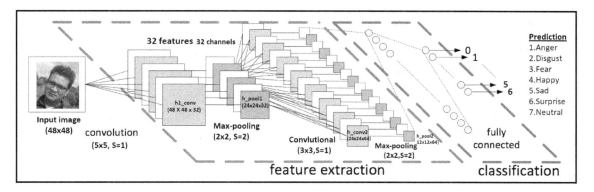

A schematic architecture of a CNN used for facial recognition

Importantly, DNNs have no prior knowledge of how the pixels are organized because they do not know that nearby pixels are close. CNNs embed this prior knowledge using lower layers by using feature maps in small areas of the image, while the higher layers combine lower-level features into larger features.

This works well with most of the natural images, giving CNNs a decisive head start over DNNs. The output from each convolutional layer is a set of objects, called feature maps, that are generated by a single kernel filter. Then, the feature maps can be used to define a new input to the next layer. Each neuron in a CNN network produces an output, followed by an activation threshold, which is proportional to the input and not bound.

RNNs

In RNNs, connections between units form a directed cycle. The RNN architecture was originally conceived by Hochreiter and Schmidhuber in 1997. RNN architectures have standard MLPs, plus added loops so that they can exploit the powerful nonlinear mapping capabilities of the MLP. They also have some form of memory. The following diagram shows a very basic RNN that has an input layer, two recurrent layers, and an output layer:

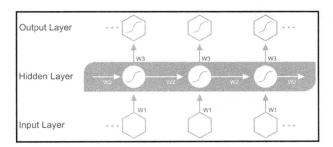

However, this basic RNN suffers from gradient vanishing and the exploding problem, and cannot model long-term dependencies. These architectures include LSTM, **gated recurrent units** (GRUs), bidirectional-LSTM, and other variants. Consequently, LSTM and GRU can overcome the drawbacks of regular RNNs: the gradient vanishing/exploding problem and long-short term dependency.

Generative adversarial networks (GANs)

Ian Goodfellow *et al.* introduced GANs in a paper named *Generative Adversarial Nets* (see more at `https://arxiv.org/abs/1406.2661v1`). The following diagram briefly shows the working principles of a GAN:

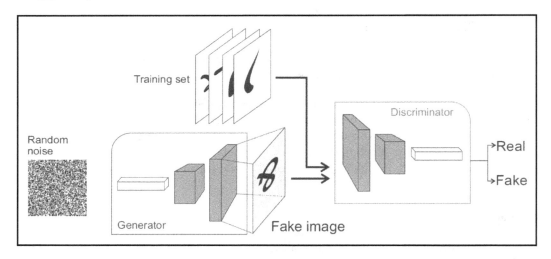

Working principles of GAN

GANs are deep neural network architectures that consist of two networks, a generator, and a discriminator, that are pitted against each other (hence the name, *adversarial*):

- The **Generator** tries to generate data samples out of a specific probability distribution and is very similar to the actual object
- The **Discriminator** will judge whether its input is coming from the original training set or from the generator part

Many DL practitioners think that GANs were one of the most important advancements because GANs can be used to mimic any distribution of data, and based on the data distribution, GANs can be taught to create robot artist images, super-resolution images, text-to-image synthesis, music, speech, and more.

For example, because of the concept of adversarial training, Facebook's AI research director, Yann LeCun, called GANs the most interesting idea in the last 10 years of ML.

Capsule networks

In CNNs, each layer understands an image at a much more granular level through a slow receptive field or max pooling operations. If the images have rotation, tilt, or very different shapes or orientation, CNNs fail to extract such spatial information and show very poor performance at image processing tasks. Even the pooling operations in CNNs cannot much help against such positional invariance. This issue in CNNs has led us to the recent advancement of CapsNet through the paper titled *Dynamic Routing Between Capsules* (see more at https://arxiv.org/abs/1710.09829) by Geoffrey Hinton *et al*:

> *"A capsule is a group of neurons whose activity vector represents the instantiation parameters of a specific type of entity, such as an object or an object part."*

Unlike a regular DNN, where we keep on adding layers, in CapsNets, the idea is to add more layers inside a single layer. This way, a CapsNet is a nested set of neural layers. In CapsNet, the vector inputs and outputs of a capsule are computed using the routing algorithm, which iteratively transfers information and process **self-consistent field** (SCF) procedure, used in physics:

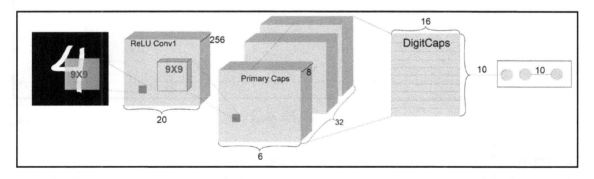

The preceding diagram shows a schematic diagram of a simple three-layer CapsNet. The length of the activity vector of each capsule in the **DigiCaps** layer indicates the presence of an instance of each class, which is used to calculate the loss.

Now that we have learned about the working principles of neural networks and the different neural network architectures, implementing something hands-on would be great. However, before that, let's take a look at some popular DL libraries and frameworks, which come with the implementation of these network architectures.

DL frameworks

There are several popular DL frameworks. Each of them comes with some pros and cons. Some of them are desktop-based and some of them are cloud-based platforms where you can deploy/run your DL applications. However, most of the libraries that are released under an open license help when people are using graphics processors, which can ultimately help in speeding up the learning process.

Such frameworks and libraries include TensorFlow, PyTorch, Keras, Deeplearning4j, H2O, and the **Microsoft Cognitive Toolkit** (**CNTK**). Even a few years back, other implementations including Theano, Caffee, and Neon were used widely. However, these are now obsolete. Since we will focus on learning in Scala, JVM-based DL libraries such as Deeplearning4j can be a reasonable choice. **Deeplearning4j** (**DL4J**) is one of the first commercial-grade, open source, distributed DL libraries that was built for Java and Scala. This also provides integrated support for Hadoop and Spark. DL4J is built for use in business environments on distributed GPUs and CPUs. DL4J aims to be cutting-edge and Plug and Play, with more convention than configuration, which allows for fast prototyping for non-researchers. The following diagram shows last year's Google Trends, illustrating how popular TensorFlow is:

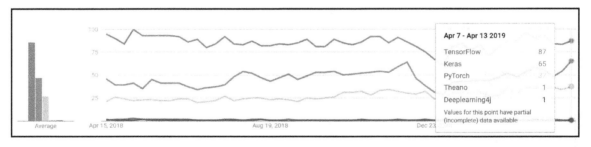

Trends of different DL frameworks—TensorFlow and Keras are dominating the most; however, Theano is losing its popularity; on the other hand, Deeplearning4j is emerging for JVM

Its numerous libraries can be integrated with DL4J and will make your JVM experience easier, regardless of whether you are developing your ML application in Java or Scala. Similar to NumPy for JVM, ND4J comes up with basic operations of linear algebra (matrix creation, addition, and multiplication). However, ND4S is a scientific computing library for linear algebra and matrix manipulation. It also provides n-dimensional arrays for JVM-based languages.

Apart from the preceding libraries, there are some recent initiatives for DL on the cloud. The idea is to bring DL capability to big data with millions of billions of data points and high dimensional data. For example, **Amazon Web Services** (**AWS**), Microsoft Azure, Google Cloud Platform, and **NVIDIA GPU Cloud** (**NGC**) all offer machine and DL services that are native to their public clouds.

In October 2017, AWS released **Deep Learning AMIs** (**DLAMIs**) for **Amazon Elastic Compute Cloud** (**Amazon EC2**) P3 instances. These AMIs come pre-installed with DL frameworks, such as TensorFlow, Gluon, and Apache MXNet, which are optimized for the NVIDIA Volta V100 GPUs within Amazon EC2 P3 instances. The DL service currently offers three types of AMIs: Conda AMI, Base AMI, and AMI with Source Code.

The CNTK is Azure's open source, DL service. Similar to AWS' offering, it focuses on tools that can help developers build and deploy DL applications. The toolkit is installed in Python 2.7, in the root environment. Azure also provides a model gallery that includes resources, such as code samples, to help enterprises get started with the service.

On the other hand, NGC empowers AI scientists and researchers with GPU-accelerated containers (see `https://www.nvidia.com/en-us/data-center/gpu-cloud-computing/`). The NGC features containerized deep learning frameworks such as TensorFlow, PyTorch, MXNet, and more that are tuned, tested, and certified by NVIDIA to run on the latest NVIDIA GPUs on participating cloud-service providers. Nevertheless, there are also third-party services available through their respective marketplaces.

Now that you know the working principles of neural network architectures and have seen a brief overview on available DL frameworks for implementing DL solutions, let's move on to the next section for some hands-on learning.

Getting started with learning

Large-scale cancer genomics data often comes in multi-platform and heterogeneous forms. These datasets impose great challenges in terms of the bioinformatics approach and computational algorithms. Numerous researchers have proposed to utilize this data to overcome several challenges, using classical ML algorithms as either the primary subject or a supporting element for cancer diagnosis and prognosis.

Description of the dataset

Genomics data covers all data related to DNA on living things. Although in this thesis we will also use other types of data, as such as transcriptomic data (RNA and miRNA), for convenience purposes, all data will be termed as genomics data. Research on human genetics made a huge breakthrough in recent years due to the success of the **Human Genome Project (HGP)** (1984-2000) on sequencing the full sequence of human DNA. Now, let's see what a real-life dataset looks like that can be used for our purposes. We will be using the *gene expression cancer RNA-Seq* dataset, which can be downloaded from the UCI ML repository (see `https://archive.ics.uci.edu/ml/datasets/ gene+expression+cancer+RNA-Seq` for more information).

This dataset is a random subset of another dataset that was reported in the following paper: Weinstein, John N., *et al. The cancer genome atlas pan-cancer analysis project*. Nature Genetics 45.10 (2013): 1113-1120. The name of the project is The Pan-Cancer analysis project. It assembled data from thousands of patients with primary tumors occurring in different sites of the body. It covered 12 tumor types, including the following:

- **Glioblastoma multiforme (GBM)**
- **Lymphoblastic acute myeloid leukemia (AML)**
- **Head and neck squamous carcinoma (HNSC)**
- **Lung adenocarcinoma (LUAD)**
- **Lung squamous carcinoma (LUSC)**
- **Breast carcinoma (BRCA)**
- **Kidney renal clear cell carcinoma (KIRC)**
- **Ovarian carcinoma (OV)**
- **Bladder carcinoma (BLCA)**
- **Colon adenocarcinoma (COAD)**
- **Uterine cervical and endometrial carcinoma (UCEC)**
- **Rectal adenocarcinoma (READ)**

This collection of data is a part of the RNA-Seq (HiSeq) PANCAN dataset. It is a random extraction of gene expressions of patients that have different types of tumors: BRCA, KIRC, COAD, LUAD, and PRAD.

This dataset is a random collection of cancer patients from 801 patients, each having 20,531 attributes. Samples (`instances`) are stored row-wise. Variables (`attributes`) of each sample are RNA-Seq gene expression levels measured by the Illumina HiSeq platform. A dummy name (`gene_XX`) is provided for each attribute. The attributes are ordered consistently with the original submission. For example, `gene_1` on `sample_0` is significantly and differentially expressed with a value of `2.01720929003`.

When you download the dataset, you will see that there are two CSV files:

- `data.csv`: Contains the gene expression data of each sample
- `labels.csv`: The labels associated with each sample

Let's take a look at the processed dataset. Note that we will only look at a few select features considering the high dimensionality in the following screenshot, where the first column represents sample IDs (that is, anonymous patient IDs). The rest of the columns represent how a certain gene expression occurs in the tumor samples of the patients:

```
|       id|gene_0|        gene_1|       gene_2|       gene_3|       gene_4|
+---------+------+--------------+-------------+-------------+-------------+
| sample_0|   0.0| 2.01720929003|3.26552691165|5.47848651208|10.4319989607|
| sample_1|   0.0|0.592732094867|1.58842082049|7.58615673813|9.62301085621|
| sample_2|   0.0| 3.5117589779|4.32719871937|6.88178695937|9.87072997113|
| sample_3|   0.0| 3.66361787431|4.50764877794|6.65906827484|10.1961840717|
| sample_4|   0.0| 2.65574107476|2.82154695883|6.53945352515|9.73826456185|
| sample_5|   0.0| 3.46785331372|3.58191760772|6.62024328973|9.70682924127|
| sample_6|   0.0|  1.224966365|1.69117679681|6.57200741498|9.64051067136|
| sample_7|   0.0| 2.85485342652|1.75047787844|7.22672044861|9.75869126501|
| sample_8|   0.0| 3.99212487426|2.77273024777|6.54669231412|10.4882518866|
| sample_9|   0.0| 3.64249364243|4.42355800269|6.84951144203|9.46446610892|
|sample_10|   0.0| 3.49207108711|  3.5533727921|7.15170663424|10.2534456958|
|sample_11|   0.0| 2.94118144936|2.66327629754|6.56168966691|9.37629255419|
|sample_12|   0.0| 3.9703475182|2.36429227014|7.1454431001|9.24060531982|
|sample_13|   0.0| 1.5510483733|3.52984592804| 6.3268249381|10.6338489327|
|sample_14|   0.0| 1.9648421858|2.18301003676|6.59683230199|10.2481410545|
|sample_15|   0.0| 2.90137860229|3.68536833781|6.66966460873|9.99909803371|
|sample_16|   0.0| 3.4609128992|3.61847360308|5.66104837265|9.73121719013|
|sample_17|   0.0| 3.00451936963|3.00717755732|6.52420475302|9.06266141243|
|sample_18|   0.0| 1.54146527849|2.54153961931|6.84325527996|9.44446829832|
|sample_19|   0.0| 4.16758272913|3.84138948407|6.97612301373|9.98225207842|
+---------+------+--------------+-------------+-------------+-------------+
only showing top 20 rows
```

Now, look at the labels in the following table. Here, the `id` column contains the sample IDs and the `Class` column represents the cancer labels:

```
+--------+-----+
|      id|Class|
+--------+-----+
|sample_0| PRAD|
|sample_1| LUAD|
|sample_2| PRAD|
|sample_3| PRAD|
|sample_4| BRCA|
|sample_5| PRAD|
|sample_6| KIRC|
|sample_7| PRAD|
|sample_8| BRCA|
|sample_9| PRAD|
+--------+-----+
only showing top 10 rows
```

Now, you can imagine why I have chosen this dataset. Although we will not have many samples, the dataset is still highly dimensional. In addition, this type of high dimensional dataset is very suitable for applying a DL algorithm. Therefore, if the features and labels are given, can we classify these samples based on features and the ground truth? Why not? We will try to solve this problem with the DL4J library. First, we have to configure our programming environment so that we can write our code.

Preparing the programming environment

In this section, we will discuss how to configure DL4J, ND4s, Spark, and ND4J before getting started with the coding. The following are the prerequisites that you must take into account when working with DL4J:

- Java 1.8+ (64-bit only)
- Apache Maven for an automated build and dependency manager
- IntelliJ IDEA or Eclipse IDE
- Git for version control and CI/CD

The following libraries can be integrated with DJ4J to enhance your JVM experience while developing your ML applications:

- **DL4J**: The core neural network framework, which comes with many DL architectures and underlying functionalities.

- **ND4J**: Can be considered as the NumPy of JVM. It comes with some basic operations of linear algebra. Examples are matrix creation, addition, and multiplication.
- **DataVec**: This library enables ETL operations while performing feature engineering.
- **JavaCPP**: This library acts as the bridge between Java and Native C++.
- **Arbiter**: This library provides basic evaluation functionalities for the DL algorithms.
- **RL4J**: Deep reinforcement learning for the JVM.
- **ND4S**: This is a scientific computing library, and it also supports n-dimensional arrays for JVM-based languages.

If you are using Maven on your preferred IDE, let's define the project properties to mention these versions in the `pom.xml` file:

```
<properties>
        <project.build.sourceEncoding>UTF-8</project.build.sourceEncoding>
        <jdk.version>1.8</jdk.version>
        <spark.version>2.2.0</spark.version>
        <nd4j.version>1.0.0-alpha</nd4j.version>
        <dl4j.version>1.0.0-alpha</dl4j.version>
        <datavec.version>1.0.0-alpha</datavec.version>
        <arbiter.version>1.0.0-alpha</arbiter.version>
        <logback.version>1.2.3</logback.version>
</properties>
```

Then, use all the dependencies required for DL4J, ND4S, and ND4J, as shown in the `pom.xml` file. By the way, DL4J comes with Spark 2.1.0. Additionally, if a native system BLAS is not configured on your machine, ND4J's performance will be reduced. You will experience the following warning once you execute any simple code written in Scala:

```
************************************************************************
WARNING: COULD NOT LOAD NATIVE SYSTEM BLAS
ND4J performance WILL be reduced
************************************************************************
```

However, installing and configuring a BLAS, such as OpenBLAS or IntelMKL, is not that difficult; you can invest some time and do it. Refer to the following URL for further details:

`http://nd4j.org/getstarted.html#open`

Well done! Our programming environment is ready for simple DL application development. Now, it's time to get our hands dirty with some sample code.

Preprocessing

Since we do not have any unlabeled data, I would like to select some samples randomly for testing. One more thing to note is that features and labels come in two separate files. Therefore, we can perform the necessary preprocessing and then join them together so that our preprocessed data will have features and labels together.

Then, the rest of the data will be used for training. Finally, we'll save the training and testing sets in a separate CSV file to be used later on. Follow these steps to get started:

1. First, let's load the samples and see the statistics. Here, we use the `read()` method of Spark, but specify the necessary options and format too:

```scala
val data = spark.read.option("maxColumns",
25000).format("com.databricks.spark.csv")
        .option("header", "true") // Use first line of all files as
header
        .option("inferSchema", "true") // Automatically infer data
types
        .load("TCGA-PANCAN/TCGA-PANCAN-HiSeq-801x20531/data.csv");//
set this path accordingly
```

2. Then, we will see some related statistics, such as the number of features and the number of samples:

```scala
val numFeatures = data.columns.length
val numSamples = data.count()
println("Number of features: " + numFeatures)
println("Number of samples: " + numSamples)
```

Therefore, there are 801 samples from 801 distinct patients and the dataset is too high in dimensions, since it has 20532 features:

```
Number of features: 20532
Number of samples: 801
```

3. In addition, since the `id` column represents only the patient's anonymous ID, so we can simply drop it:

```scala
val numericDF = data.drop("id") // now 20531 features left
```

4. Then, we load the labels using the `read()` method of Spark and also specify the necessary options and format:

```
val labels = spark.read.format("com.databricks.spark.csv")
        .option("header", "true")
        .option("inferSchema", "true")
        .load("TCGA-PANCAN/TCGA-PANCAN-HiSeq-801x20531/labels.csv")
labels.show(10)
```

We have already seen what the label DataFrame looks like. We will skip the `id`. However, the `Class` column is categorical. As we mentioned previously, DL4J does not support categorical labels that need to be predicted. Therefore, we have to convert it into a numeric format (an integer, to be more specific); for that, I would use `StringIndexer()` from Spark:

1. First, we create a `StringIndexer()`, we apply the index operation to the `Class` column, and rename it `label`. Additionally, we `skip` null entries:

```
val indexer = new StringIndexer().setInputCol("Class")
                .setOutputCol("label")
                .setHandleInvalid("skip"); // skip null/invalid
values
```

2. Then, we perform the indexing operation by calling the `fit()` and `transform()` operations, as follows:

```
val indexedDF = indexer.fit(labels).transform(labels)
                    .select(col("label")
                    .cast(DataTypes.IntegerType)); // casting
data types to integer
```

3. Now, let's take a look at the indexed DataFrame:

```
indexedDF.show()
```

The preceding line of code should convert the `label` column in numeric format:

```
+-----+
|label|
+-----+
|    3|
|    2|
|    3|
|    3|
|    0|
|    3|
|    1|
|    3|
|    0|
|    3|
+-----+
only showing top 10 rows
```

4. Fantastic! Now, all the columns (including features and labels) are numeric. Thus, we can join both features and labels into a single DataFrame. For that, we can use the `join()` method from Spark, as follows:

```
val combinedDF = numericDF.join(indexedDF)
```

5. Now, we can generate both the training and test sets by randomly splitting `combinedDF`, as follows:

```
val splits = combinedDF.randomSplit(Array(0.7, 0.3), 12345L) //70%
for training, 30% for testing
val trainingDF = splits(0)
val testDF = splits(1)
```

6. Now, let's see the `count` of samples in each set:

```
println(trainingDF.count())// number of samples in training set
println(testDF.count())// number of samples in test set
```

7. There should be 561 samples in the training set and 240 samples in the test set. Finally, we save them in separate CSV files to be used later on:

```
trainingDF.coalesce(1).write
    .format("com.databricks.spark.csv")
    .option("header", "false")
    .option("delimiter", ",")
    .save("output/TCGA_train.csv")
```

```
testDF.coalesce(1).write
        .format("com.databricks.spark.csv")
        .option("header", "false")
        .option("delimiter", ",")
        .save("output/TCGA_test.csv")
```

8. Now that we have the training and test sets, we can train the network with the training set and evaluate the model with the test set.

 Spark will generate CSV files under the output folder, under the project root. However, you might see a very different name. I suggest that you to rename them TCGA_train.csv and TCGA_test.csv for the training and test sets, respectively.

Considering the high dimensionality, I would rather try a better network such as LSTM, which is an improved variant of RNN. At this point, some contextual information about LSTM would be helpful to grasp this idea, and will be provided after the following section.

Dataset preparation

In the previous section, we prepared the training and test sets. However, we need to put some extra effort into making them consumable by DL4J. To be more specific, DL4J expects the training data in numeric format and the last column to be the label column. The remaining data should be features.

We will now try to prepare our training and test sets like that. First, we will find the files where we saved the training and test sets:

```
// Show data paths
val trainPath = "TCGA-PANCAN/TCGA_train.csv"
val testPath = "TCGA-PANCAN/TCGA_test.csv"
```

Then, we will define the required parameters, such as the number of features, number of classes, and batch size. Here, I am using 128 as the batchSize, but you can adjust it accordingly:

```
// Preparing training and test set.
val labelIndex = 20531
val numClasses = 5
val batchSize = 128
```

This dataset is used for training:

```
val trainingDataIt: DataSetIterator = readCSVDataset(trainPath, batchSize,
labelIndex, numClasses)
```

This is the data we want to classify:

```
val testDataIt: DataSetIterator = readCSVDataset(testPath, batchSize,
labelIndex, numClasses)
```

As you can see from the preceding two lines of code, `readCSVDataset()` is basically a wrapper that reads the data in CSV format, and then the `RecordReaderDataSetIterator()` method converts the record reader into a dataset iterator.

LSTM network construction

Creating a neural network with DL4J starts with `MultiLayerConfiguration`, which organizes network layers and their hyperparameters. Then, the created layers are added using the `NeuralNetConfiguration.Builder()` interface. As shown in the following diagram, the LSTM network consists of five layers: an input layer, which is followed by three LSTM layers. The last layer is an RNN layer, which is also the output layer in this case:

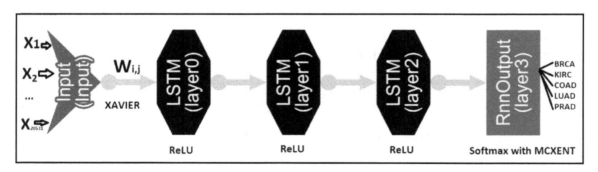

An LSTM network for cancer type prediction, which takes 20,531 features and fixed bias (that is, 1) and generates multi-class outputs

To create LSTM layers, DL4J provides the implementation of an LSTM class. However, before we start creating layers for the network, let's define some hyperparameters, such as the number of input/hidden/output nodes (neurons):

```
// Network hyperparameters
val numInputs = labelIndex
val numOutputs = numClasses
val numHiddenNodes = 5000
```

We then create the network by specifying layers. The first, second, and third layers are LSTM layers. The last layer is an RNN layer. For all the hidden LSTM layers, we specify the number of input and output units, and we use ReLU as the activation function. However, since it's a multiclass classification problem, we use SOFTMAX as the activation function for the output layer, with MCXNET as the loss function:

```
//First LSTM layer
val layer_0 = new LSTM.Builder()
    .nIn(numInputs)
    .nOut(numHiddenNodes)
    .activation(Activation.RELU)
    .build()

//Second LSTM layer
val layer_1 = new LSTM.Builder()
    .nIn(numHiddenNodes)
    .nOut(numHiddenNodes)
    .activation(Activation.RELU)
    .build()

//Third LSTM layer
val layer_2 = new LSTM.Builder()
    .nIn(numHiddenNodes)
    .nOut(numHiddenNodes)
    .activation(Activation.RELU)
    .build()

//RNN output layer
val layer_3 = new RnnOutputLayer.Builder()
    .activation(Activation.SOFTMAX)
    .lossFunction(LossFunction.MCXENT)
    .nIn(numHiddenNodes)
    .nOut(numOutputs)
    .build()
```

In the preceding code block, the softmax activation function gives a probability distribution over classes, and MCXENT is the cross-entropy loss function in a multiclass classification setting.

Then, with DL4J, we add the layers we created earlier using the `NeuralNetConfiguration.Builder()` interface. First, we add all the LSTM layers, which are followed by the final RNN output layer:

```
//Create network configuration and conduct network training
val LSTMconf: MultiLayerConfiguration = new
NeuralNetConfiguration.Builder()
    .seed(seed) //Random number generator seed for improved
repeatability. Optional.
    .optimizationAlgo(OptimizationAlgorithm.STOCHASTIC_GRADIENT_DESCENT)
    .weightInit(WeightInit.XAVIER)
    .updater(new Adam(5e-3))
    .l2(1e-5)
    .list()
        .layer(0, layer_0)
        .layer(1, layer_1)
        .layer(2, layer_2)
        .layer(3, layer_3)
    .pretrain(false).backprop(true).build()
```

In the preceding code block, we used SGD as the optimizer, which tries to optimize the MCXNET loss function. Then, we initialize the network weight using XAVIER, and Adam acts as the network updater with SGD. Finally, we initialize a multilayer network using the preceding multilayer configuration:

```
val model: MultiLayerNetwork = new MultiLayerNetwork(LSTMconf)
model.init()
```

Additionally, we can inspect the number of hyperparameters across layers and in the whole network. Typically, this type of network has a lot of hyperparameters. Let's print the number of parameters in the network (and for each layer):

```
//print the score with every 1 iteration
model.setListeners(new ScoreIterationListener(1))

//Print the number of parameters in the network (and for each layer)
val layers = model.getLayers()
var totalNumParams = 0
var i = 0

for (i <- 0 to layers.length-1) {
    val nParams = layers(i).numParams()
    println("Number of parameters in layer " + i + ": " + nParams)
    totalNumParams = totalNumParams + nParams
}
println("Total number of network parameters: " + totalNumParams)
```

The output of the preceding code is as follows:

```
Number of parameters in layer 0: 510640000
Number of parameters in layer 1: 200020000
Number of parameters in layer 2: 200020000
Number of parameters in layer 3: 25005
Total number of network parameters: 910705005
```

As I stated previously, our network has 910 million parameters, which is huge. This also poses a great challenge while tuning hyperparameters.

Network training

First, we will create a `MultiLayerNetwork` using the preceding `MultiLayerConfiguration`. Then, we will initialize the network and start the training on the training set:

```
var j = 0
println("Train model....")
for (j <- 0 to numEpochs-1) {
    model.fit(trainingDataIt)
```

Finally, we also specify that we do not need to do any pre-training (which is typically needed in DBN or stacked autoencoders).

Evaluating the model

Once the training has been completed, the next task is to evaluate the model, which we'll do on the test set here. For the evaluation, we will be using the `Evaluation()` method. This method creates an evaluation object with five possible classes.

First, let's iterate the evaluation over every test sample and get the network's prediction from the trained model. Finally, the `eval()` method checks the prediction against the true class:

```
println("Evaluate model....")
val eval: Evaluation = new Evaluation(5) //create an evaluation object with
5 possible classes
while (testDataIt.hasNext()) {
    val next:DataSet = testDataIt.next()
    val output:INDArray  = model.output(next.getFeatureMatrix()) //get
the networks prediction
    eval.eval(next.getLabels(), output) //check the prediction against
```

```
the true class
    }
println(eval.stats())
println("***************Example finished********************")
  }
```

The following is the output:

```
===========================Scores==========================================
    # of classes:    5
    Accuracy:        0.9900
    Precision:       0.9952
    Recall:          0.9824
    F1 Score:        0.9886
 Precision, recall & F1: macro-averaged (equally weighted avg. of 5
classes)
    ====================================================================
    ***************Example finished*******************
```

Wow! Unbelievable! Our LSTM network has accurately classified the samples. Finally, let's see how the classifier predicts across each class:

```
Actual label 0 predicted by the model as 0: 82 times
Actual label 1 predicted by the model as 0: 1 times
Actual label 1 predicted by the model as 1: 17 times
Actual label 2 predicted by the model as 2: 35 times
Actual label 3 predicted by the model as 0: 1 times
Actual label 3 predicted by the model as 3: 30 times
```

The predictive accuracy for cancer type prediction using LSTM is suspiciously higher, isn't it? Did our model underfit? Did our model overfit?

Observing the training using Deeplearning4j UI

As our accuracy is suspiciously higher, we can observe how the training went. Yes, there are ways to find out if it went through overfitting, since we can observe the training, validation, and test losses on the DL4J UI. However, I won't discuss the details here. Take a look at `https://deeplearning4j.org/docs/latest/deeplearning4j-nn-visualization` for more information on how to do this.

Summary

In this chapter, we saw how to classify cancer patients on the basis of tumor types from a very high-dimensional gene expression dataset curated from TCGA. Our LSTM architecture managed to achieve 99% accuracy, which is outstanding. Nevertheless, we discussed many aspects of DL4J, which will be helpful in upcoming chapters. Finally, we saw answers to some frequent questions related to this project, LSTM networks, and DL4J hyperparameters/network tuning.

This is, more or less, the end of our little journey in developing ML projects using Scala and different open source frameworks. Throughout these chapters, I have tried to provide you with several examples of how to use these wonderful technologies efficiently for developing ML projects. While writing this book, I had to keep many constraints in my mind; for example, the page count, API availability, and of course, my expertise.

However, overall, I tried to make the book simple by avoiding unnecessary details on the theory, as you can read about that in many books, blogs, and websites. I will also keep the code of this book updated on the GitHub repository at `https://github.com/PacktPublishing/Machine-Learning-with-Scala-Quick-Start-Guide`. Feel free to open a new issue or any pull request to improve the code and stay tuned.

Nevertheless, I'll upload the solution to each chapter as Zeppelin notebooks so that you can run the code interactively. By the way, Zeppelin is a web-based notebook that enables data-driven, interactive data analytics, and collaborative documents with SQL and Scala. Once you have configured Zeppelin on your preferred platform, you can download the notebook from the GitHub repository, import them into Zeppelin, and get going. For more detail, you can take a look at `https://zeppelin.apache.org/`.

Other Books You May Enjoy

If you enjoyed this book, you may be interested in these other books by Packt:

Scala Machine Learning Projects
Md. Rezaul Karim

ISBN: 978-1-78847-904-2

- Apply advanced regression techniques to boost the performance of predictive models
- Use different classification algorithms for business analytics
- Generate trading strategies for Bitcoin and stock trading using ensemble techniques
- Train Deep Neural Networks (DNN) using H2O and Spark ML
- Utilize NLP to build scalable machine learning models
- Learn how to apply reinforcement learning algorithms such as Q-learning for developing ML application
- Learn how to use autoencoders to develop a fraud detection application
- Implement LSTM and CNN models using DeepLearning4j and MXNet

Scala and Spark for Big Data Analytics
Md. Rezaul Karim, Sridhar Alla

ISBN: 978-1-78528-084-9

- Understand object-oriented & functional programming concepts of Scala
- In-depth understanding of Scala collection APIs
- Work with RDD and DataFrame to learn Spark's core abstractions
- Analysing structured and unstructured data using SparkSQL and GraphX
- Scalable and fault-tolerant streaming application development using Spark structured streaming
- Learn machine-learning best practices for classification, regression, dimensionality reduction, and recommendation system to build predictive models with widely used algorithms in Spark MLlib & ML
- Build clustering models to cluster a vast amount of data
- Understand tuning, debugging, and monitoring Spark applications
- Deploy Spark applications on real clusters in Standalone, Mesos, and YARN

Leave a review - let other readers know what you think

Please share your thoughts on this book with others by leaving a review on the site that you bought it from. If you purchased the book from Amazon, please leave us an honest review on this book's Amazon page. This is vital so that other potential readers can see and use your unbiased opinion to make purchasing decisions, we can understand what our customers think about our products, and our authors can see your feedback on the title that they have worked with Packt to create. It will only take a few minutes of your time, but is valuable to other potential customers, our authors, and Packt. Thank you!

Index

www.ingramcontent.com/pod-product-compliance
Lightning Source LLC
Chambersburg PA
CBHW080525060326
40690CB00022B/5026